Programming for Optimal Decisions

Selected Readings in Mathematical Programming Techniques for Management Problems

Edited by P. G. Moore and S. D. Hodges

Penguin Books

Penguin Books Ltd, Harmondsworth,
Middlesex, England
Penguin Books Inc., 7110 Ambassador Road,
Baltimore, Md 21207, U.S.A.
Penguin Books Australia Ltd, Ringwood,
Victoria, Australia

First published 1970

Made and printed in Great Britain by
C. Nicholls & Company Ltd
Set in Monotype Times

Contents

Part One Introduction

The problem of the efficient allocation of resources faces all organizations in one or more forms, and the techniques of mathematical programming (of which linear programming forms a part) provide the most universal tool yet devised for tackling it. This book attempts to bring together a number of papers to illustrate the wide range of applications which have been made of the many techniques now available.

Historically, the methods of mathematical programming are quite recent – for when in 1945 George Stigler considered an example of the diet problem (of finding the cheapest diet, from among seventy-seven foods, satisfying nine different nutritional requirements) he had to content himself with a near-optimal solution obtained by a 'trial and error' procedure. He wrote: 'There does not appear to be any direct method of finding the minimum of a linear function subject to linear conditions', which was indeed true at that time. It was another two years before in 1947 George Dantzig was to discover the Simplex method, and it was not published until 1951.

The 'standard form' of linear programming problem may be stated mathematically as follows:

Choose the quantities

$$x_j \geqslant O \quad (j = 1, \ldots, n)$$

to *maximize*

$$P = \sum_{j=1}^{n} c_j x_j,$$

subject to the constraints

$$\sum_{j=1}^{n} a_{ij} x_j \leqslant b_i \quad (i=1, \ldots, m).$$

This can often be interpreted as a profit maximization situation where n production activities are pursued at levels x_j which have to be decided upon, subject to only limited amounts of m resources being available. Each unit of the jth activity yields a return c_j and uses an amount a_{ij} of the ith resource. The Simplex method initially obtains a feasible solution (i.e., one satisfying the constraint equations, but not necessarily the optimum) to the linear programme (L P). This is modified in a series of iterations until no further improvement is possible. The method was first successfully used on a computer in 1952 and today, of course, many efficient computer codes for the Simplex method are available.

One type of linear programme of special importance is the transportation problem first formulated independently by Hitchcock in 1941 and Koopmans in 1947. This is the problem of minimizing the cost of shipping quantities of a product from several origins to several destinations. Mathematically it is stated as:

choose

$$x_{ij} \geqslant 0 \quad (i = 1, \ldots, m; \quad j = 1, \ldots, n)$$

to *minimize*

$$\sum_{i=1}^{m} \sum_{j=1}^{n} c_{ij} x_{ij},$$

subject to

$$\sum_{j=1}^{n} x_{ij} = a_i \quad (i = 1, \ldots, m)$$

and

$$\sum_{i=1}^{m} x_{ij} = b_j \quad (j = 1, \ldots, n).$$

Here

$$\sum_{i=1}^{m} a_i = \sum_{j=1}^{n} b_j$$

is the total quantity to be shipped, origin i is to supply an amount a_i, and destination j to receive b_j. The unit cost of shipment from origin i to destination j is c_{ij}. An especially simple technique was developed by Dantzig in 1951 to solve problems of this structure, which often arise even in contexts very different from that of transportation. Where an L P can be structured in this form, either directly or as an approximation, great savings in computation may be made.

The great power and applicability of these new techniques to many management situations soon fired the imagination of managers and researchers alike, leading to a phenomenal development of the field of mathematical programming. To some extent the idea was oversold and too often problems were distorted until they could be fitted into the linear programming mould. For an LP solution to be valid, it must be possible to describe the system adequately by linear expressions whose coefficients can be determined in advance and without significant inaccuracies or random deviations. Overlooking these conditions led to useless and possibly damaging results being obtained: a further stimulus to the development of a wide range of techniques. Parametric programming allowed sensitivity analyses to be made to examine the effect of changes in some of the numerical coefficients. Faster and more accurate computer codes capable of solving larger and larger problems were developed, along with special methods for dealing efficiently with LPs of certain structures. The original basic theory was extended in many directions: in particular, to cope with non-linearity, with random variables and with the problem of finding integer solutions to LPs. While many of these problems are still only partially solved, in other cases the techniques seem to run ahead of the applications for them, and it is more and more necessary to maintain interaction between the theoretical and the practical.

The second and third parts of this book consist of papers which illustrate the application of mathematical programming in different forms to a wide variety of kinds of resource allocation. Part Two is restricted to linear programming, while Part Three illustrates the use of some forms of non-linear, probabilistic and integer programming and also a heuristic approach. These papers can be read as case studies in the allocation of resources and without any deep knowledge of the underlying theory. In order to present a fuller and more general view of some of the more recent techniques, the last part of the book contains papers of a mainly theoretical nature and may be more demanding on the reader. The first paper in the book, following this introduction is of a semi-technical nature, and should give additional perspective to mathematical programming in relation to other optimization procedures.

1 J. E. Mulligan

Basic Optimization Techniques – A Brief Survey

J. E. Mulligan, 'Basic optimization techniques – a brief survey', *Journal of Industrial Engineering*, vol. 16, 1965, pp. 192–7.

Sometimes enthusiasm for a new technique leads to a sort of mental suboptimization. For instance, one might become so intrigued with linear programming that, faced with a specific optimization problem, there would be two steps to his approach to a solution:

1. How does one structure this problem as a linear program?
2. What is the best method of solution for the resulting linear program?

Hopefully there are few people who are this rigid in their approach, but there are some who seem to be veering dangerously in this direction. Their effect on innocent bystanders is devastating. What operations research oriented person has not been hailed into his boss' office to hear something like the following. 'From what I hear and read, everybody seems to be doing well with linear programming. Why aren't we?' To such bosses it is often a revelation to hear that linear programming is one of many approaches to optimization. They are pleased to find names like dynamic programming, hill climbing, Lagrange multipliers, and simulation that represent other approaches to the same problem. They are pleased because they can now associate these names under a common heading – optimization techniques. (The name of any optimization technique could be substituted for 'linear programming' in the question which the boss raised. Linear programming is used as an example.)

To test your own position in the continuum of possible biases, consider the following problem:
'A company makes two kinds of leather belts. Belt A is a high quality belt and belt B is of lower quality. The respective profits are $0.40 and $0.30 per belt. Each belt of type A requires twice

as much time as a belt of type B, and, if all belts were of type B the company could make 1000 per day. The supply of leather is sufficient for only 800 belts per day (both A and B combined). Belt A requires a fancy buckle, and only 400 per day are available. There are only 700 buckles a day available for belt B.

Assuming that all the A belts and B belts the factory can make can be sold, how should the manager schedule production to produce the most profit? (Sasieni *et al.*, 1959, p. 236.)

The problem is stated in the usual terms, and they are shown in the glossary as follows:

Z = profit
X_1 = number of A belts produced
X_2 = number of B belts produced
X_3 = remaining A buckles
X_4 = remaining B buckles
X_5 = remaining units of leather
X_6 = remaining time units.

The objective function is now specified as:

$$Z = 0{\cdot}40X_1 + 0{\cdot}30X_2$$

and the constraint equations:

$$\begin{aligned} X_1 &\leqslant 400 \\ X_2 &\leqslant 700 \\ X_1 + X_2 &\leqslant 800 \\ 2X_1 + X_2 &\leqslant 1000. \end{aligned}$$

What is your first reaction?

An engineer (who once took *Advanced Calculus for Engineers*) might say: 'That's a *Lagrange multiplier* problem' (Sokolnikoff, 1939, p. 327). Of course he is right. Change the inequalities to equalities

$$\begin{aligned} X_1 &= 400 \\ 2X_1 + X_2 &= 1000 \\ X_1 + X_2 + X_5 &= 800 \\ X_2 + X_4 &= 700 \end{aligned}$$

and create the Lagrangian expression:

$$\begin{aligned} L = 0{\cdot}40X_1 + 0{\cdot}30X_2 &+ \lambda_1(X_1 - 400) + \lambda_2(X_2 + X_4 - 700) \\ &+ \lambda_3(X_1 + X_2 + X_5 - 800) + \lambda_4(2X_1 + X_2 - 1000). \end{aligned}$$

Differentiating with respect to each of the eight variables, setting the resulting expressions equal to zero, and solving the simultaneous equations would yield:

$$X_1 = \text{A belts produced} = 400$$
$$X_2 = \text{B belts produced} = 200$$
$$X_4 = \text{B buckles remaining} = 500$$
$$X_5 = \text{leather units remaining} = 200$$
$$\lambda_1 = +\,0{\cdot}20$$
$$\lambda_2 = 0$$
$$\lambda_3 = 0$$
$$\lambda_4 = -\,0{\cdot}30.$$

Our engineer would then point out that the optimum had not yet been found. 'The + 0·20 value for λ_1 means that $0.20 can be made by eliminating an A buckle (and hence an A belt).' He might then proceed to revise his constraint equations as follows:

$$X_1 + X_3 - 400 = 0$$
$$X_2 + X_4 - 700 = 0$$
$$X_1 + X_2 - 800 = 0$$
$$2X_1 + X_2 - 1000 = 0.$$

From this there would result a new Lagrangian expression, new simultaneous equations and the following result:

$$X_1 = 200$$
$$X_2 = 600$$
$$X_3 = 200$$
$$X_4 = 100$$
$$\lambda_1 = 0$$
$$\lambda_2 = 0$$
$$\lambda_3 = -\,0{\cdot}20$$
$$\lambda_4 = -\,0{\cdot}10.$$

The engineer would now turn to his audience and claim this was the optimum solution. 'Two hundred A belts and 600 B belts – that's the answer. The negative λ's mean that money would be lost if one less A belt or one less B belt was produced. Let's see, the profit would be $260.'

In any audience of reasonable size, there would certainly be one person who would jump up at this point and say: 'Actually this is a *linear programming* problem. Let's check your solution.' Of course this man is also right, and he might produce the diagrams (Ackoff, 1961, pp. 126–7) in Figures 1 and 2. The equal cost curve through extreme point C is furthest from the origin and yields $260 profit. The engineer thanks the linear programmer for the check but politely notes that if the profit or constraint equations turn out to be nonlinear, linear programming is in trouble. On the other hand, Lagrange multipliers could still be

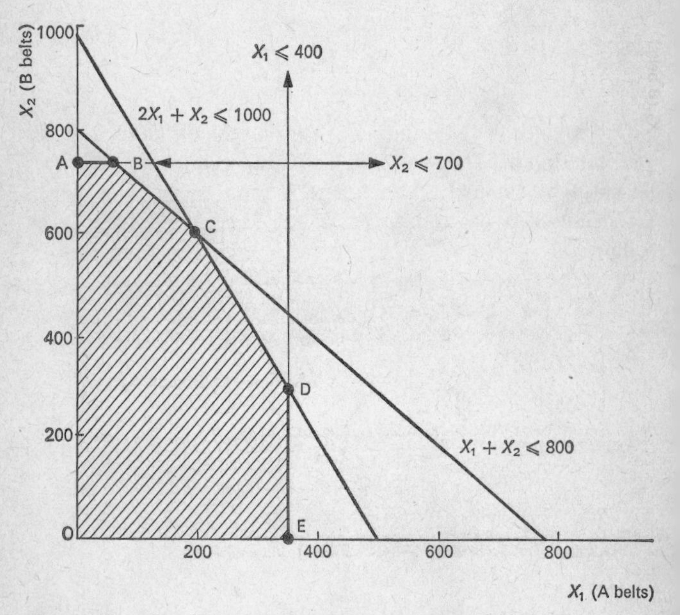

Figure 1 Functional structure linear programming solution

used. The linear programmer replies: 'There's still nonlinear programming. Seriously, I can see you taking months to come up with some multivariable, nonlinear representation of this problem, spending hours of computer time inverting a huge matrix, and finally concluding that this factory should make 199·51 A belts and 599·49 B belts for a profit of $259.999. You talk about method. How would you tell a computer to go from the first pass you made to the second pass?'

The engineer stammers a bit, but a pacifier arises and says, 'Gentlemen, gentlemen. Come now. Linear or non-linear, *dynamic programming* gives you the general method of solution. Further, when this business expands we're ready to handle new products.' He then proceeds to draw up (after Vazsonyi,1958) Tables 1 and 2. 'You can see $260 is the right answer as the problem is stated and furthermore, we're ready to take on a new belt.'

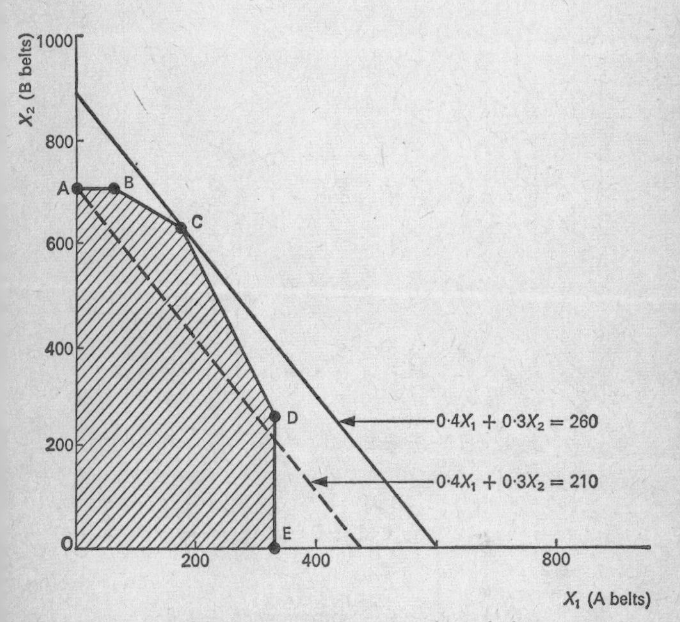

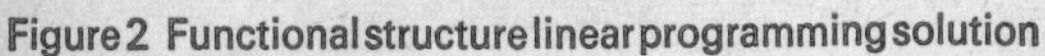
Figure 2 Functional structure linear programming solution

At this point another person might well arise and say: 'Gentlemen! Although the solutions you've suggested may be elegant, you're all off on the wrong track. Your objective of trying to fit mathematical functions to real-world problems is just plain impractical. The only way to research real-world problems is *simulation* (Ackoff, 1961, ch. 9). Logically structure the problem – identify the blocks or modules – then prepare flow charts. This way you are not trying to avoid nonlinearities nor invent them.' He then proceeds to draw up Figure 3, and, after noting that an

Table 1

Functional Structure Dynamic Programming Solution – Profit Possibilities for X_1 and X_2

		X_1 *made*	*Tenths of days producing* X_1										
			0	1	2	3	4	5	6	7	8	9	10
X_2 *made*			0	50	100	150	200	250	300	350	400	450	500
	0	0	0	20	40	60	80	100	120	140	160		
	1	100	30*	50	70	90	110	130	150	170	190		
	2	200	60*	80	100	120	140	160	180	200	220	↓	
	3	300	90*	110	130	150	170	190	210	230		$X_1=400$	
Tenths of	4	400	120*	140	160	180	200	220	240				
days	5	500	150*	170	190	210	230	250					
producing	6	600	280*	200	220	240	260*						
X_2	7	700	110*	230*	250*			→ $X_1+X_2=800$					
								→ $X_2=700$					
	8	800											
	9	900				↓							
	10	1000				$X_1+X_2=800$							

* Maxima along each diagonal

Table 2

Functional Structure Dynamic Programming Solution – Profit Possibilities X_1, X_2 and X_{new}

		X_1 *and* X_2 *made*	*Tenths of days producing* X_1 *and* X_2										
			0	1	2	3	4	5	6	7	8	9	10
X_{new} *made*			0	100	200	300	400	500	600	700	750	800	800
	0	0	0	30	60	90	120	150	180	210	230	250	260
	1	—	—	—	—	—	—	—	—	—	—	—	
	2	—	—	—	—	—	—	—	—	—	—		
	3	—	—	—	—	—	—	—	—	—			
Tenths of days producing X_{new}	4	—	—	—	—	—	—	—	—				
	5	—	—	—	—	—	—	—					
	6	—	—	—	—	—	—						
	7	—	—	—	—	—							
	8	—	—	—	—								
	9	—	—	—									
	10	—	—										

event-type model would be more efficient than a time-step model, he draws Figures 4 and 5. 'And if you can fit your logic on a computer I suppose you are going to recommend a *Monte Carlo* solution,' sneers the linear programmer. 'Then you'll run case after case and never know whether you've really found the optimum. You're just as bad if not worse than the engineer and his *n*th degree polynomials. You'll replace every term in his equations with a sub-routine.' The dynamic programmer starts to point out that his method of solution will cover this structure as well as any other. But the engineer – finally recovering his poise – advances to the board.

'You don't have to code up a modified simplex algorithm or fill in all those boxes in the dynamic programming tables or draw numbers out of a hat. There is a perfectly logical way to find optimums in all cases – derivatives will point the way to the top. And just like dynamic programming, this method applies to either structure, but it will save a lot of computation. I was using derivatives in the Lagrange multiplier solution. Let's use the *gradient method* (Wilde, 1964) to find the summit of the profit mountain structured by these flow charts. Let's start with a base point of 200 B belts.'

	Condition	Resulting A Belts	Resulting Profit	Δ
B	(100 B belts)	400	$190	
				−30
A	(200 B belts)	400	$220	
				+10
C	(300 B belts)	350	$230	

Pointing out that the gradient (Δ) shows that more B belts should be produced, he moves on to a base point of 500 and eventually, after cutting step size, comes up with Figure 6. He admits there could have been further saving. He would not have had to evaluate so many derivatives had he taken time to '*sample the profit space*' before using the gradient technique. Several samples are suggested and numerous points are added to Figure 6. The meeting then degenerates into a discussion of significance of sample design (and, of course, it is significant), but let us withdraw and consider what we, or the manager of a belt factory, can deduce from the discussion to this point.

Conclusions

These men as a body were not ready to legislate that the world is polynomial (let alone linear) or a series of flow charts. Therefore, it appears that one must make a decision about how to *structure* a given problem. But this is not the end. Having decided upon a structure, one must choose among several *methods of solution.*

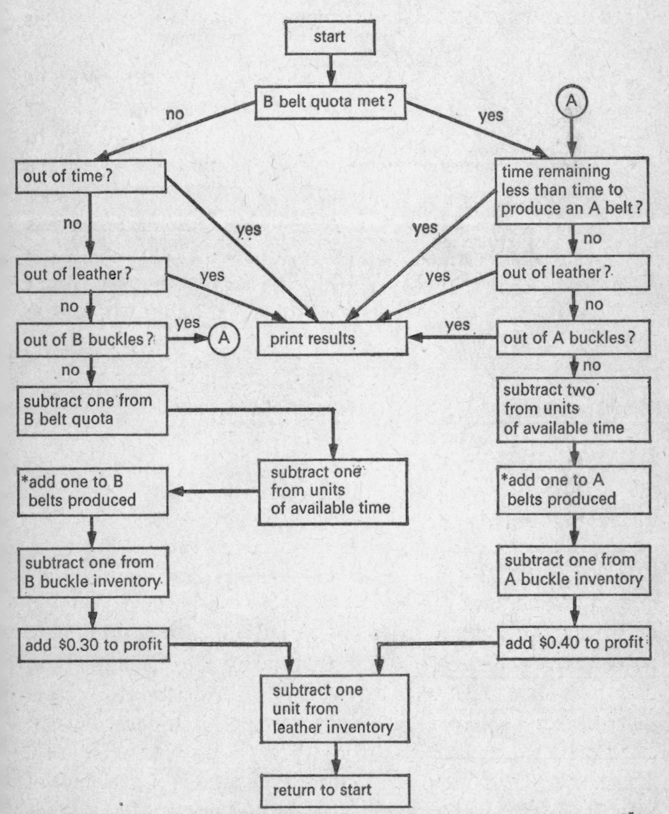

*it is these boxes that makes this a 'stepping' model. Belts are produced one at a time.

Figure 3 English language flow chart – stepping model

One might attempt to draw up a flow chart for the proper approach to an analytical effort. Figure 7 is a more comprehensive approach than the one that was dubbed rigid in the opening paragraph, but a reviewer might term this figure 'a stronger and more rigid approach than is intended by the author.' The figure does however serve to summarize some of the salient points

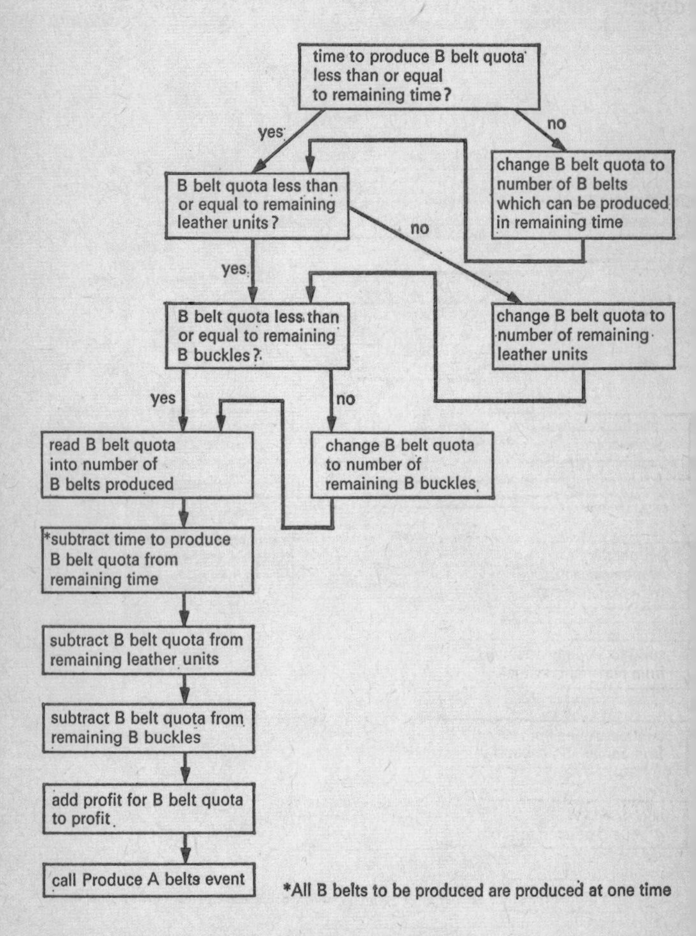

Figure 4 English language flowchart – event type model (B belts)

brought out in the discussion of various assaults on the belt factory problem. First, it is important to note the two ways of structuring optimization problems for mathematical analysis, *functional* structuring and *logical* structuring. Although both of these methods of structuring share many problems (data gathering, programming, check-out, and so forth), the two methods are different and each has basic advantages and disadvantages.

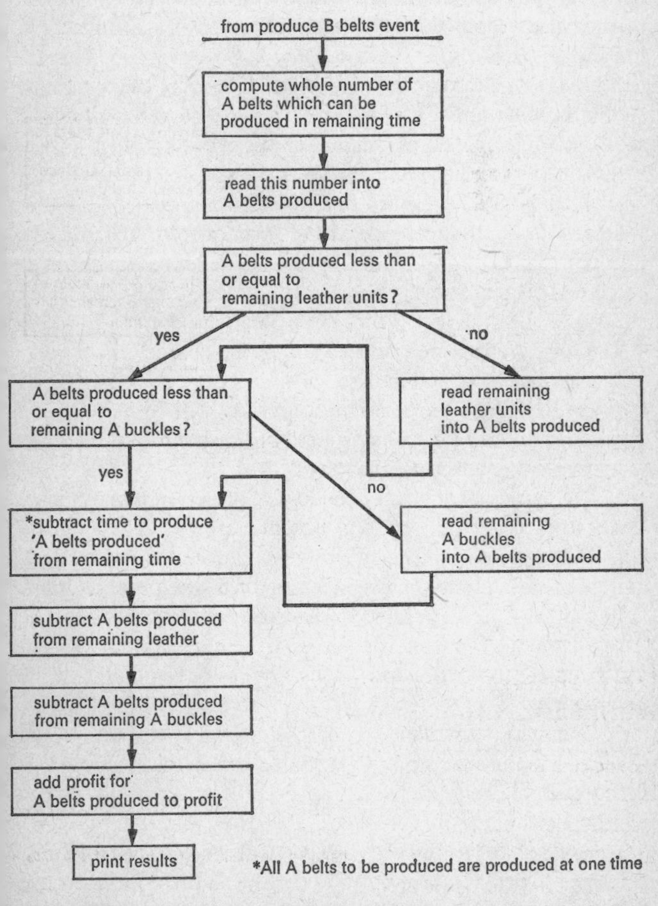

Figure 5 English language flow chart – event type model (A belts)

The essential points concerning functional structures are as follows:

1. The functional structure uses mathematical functions or equations to describe the process or system being studied. Each term describes a contribution of one or more of the parameters involved.

2. The basic advantage of functional structuring is the opportunity to employ conventional, although perhaps complicated, mathematical tools (that is, differentiation, Lagrange multipliers, linear programming) in the attempt to solve a problem. This advantage is lost if the equations required to describe the process or system become too complex. In this latter case, the methods of solution (dynamic programming, statistical analysis of experiments or hill climbing techniques) are the same methods of solution which would be applied to a logical structure.

3. The basic disadvantage of the functional approach is the time-consuming effort involved in finding the proper functions or equations to realistically describe the process or system being studied.

The logical structures have the following characteristics:

1. The logical structures uses flow charts or structure tables (Kavanagh, 1963) to describe the process or systems being studied, each block or module describing the contribution of one or more of the parameters involved.

2. The basic advantage of the logical structure is the opportunity to describe the process or system more precisely. The burden of determining the equations which link a large number of variables – of determining significance of higher order or interacting terms – has been lifted by the choice of a logical structure. This advantage is lost if the flow charts or structure tables are too extensive to fit the constraints (physical or economic) of the computer.

3. The basic disadvantage of the logical approach is the time-consuming method of solution to which this approach is committed, namely, statistical analysis of experiments or hill climbing techniques.

It may be hard for us with our particular bents (be they functional or logical) to face-up to the signal that comes from this discussion – but the signal is clear. It is possible to push one's

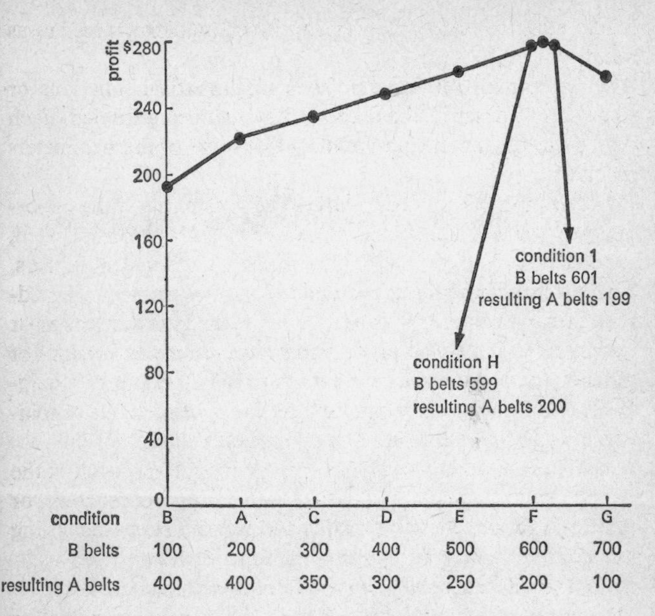

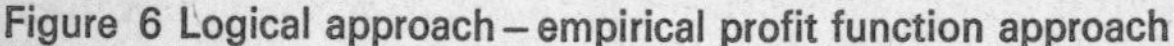

Figure 6 Logical approach – empirical profit function approach

'pet' procedure far beyond the bounds of economic utility unless control is exercised and, in so doing, more promising approaches may be ignored. There is a need for a management or self-disciplinary approach to our work. One wants to be particularly concerned with the potentially expensive efforts (the A items), less concerned with the moderately expensive efforts (the B items) and least concerned with the inexpensive efforts (the C items). The symptoms that allow one to discriminate between A, B and C items were noted in the preceding discussion and are summarized as follows:

	Functional approach with constraints	*Logical (simulation) approach*
A Symptoms	Many variables.	Many blocks (modules).
	Many constraints.	Much detail in all blocks.
	Many terms of second or higher degree.	

	Functional approach with constraints	*Logical (simulation) approach*
B Symptoms	Several variables.	Several blocks.
	Several constraints.	Several in great detail.
	Some terms of second degree.	
C Symptoms	Few variables.	Few blocks.
	Few constraints.	A few in moderate detail.
	Most terms first degree.	

In no way is the foregoing tabulation intended to imply that the A analytical efforts should be avoided. This would be like saying there should be no A components in an assembly. The point is that our sponsors cannot afford to have too many A and B efforts in their research programs. The C efforts on the other hand, being inexpensive, should be widely employed and immediate profit taken.

The table of symptoms hints as to one's approach, once the nature (A, B, or C) of a particular research effort has been diagnosed. If an A or B effort using the functional approach is embarked upon, someone should be continually asking if all the constraints and variables included in the structure are justified in the light of the objective. Is there redundancy in the constraint equations? Is it necessary to include this quadratic term? Is the degree across terms consistent? The questions in the case of the logical approach are similar. Is this particular block (module) necessary? Must there be such detailed coverage of a particular decision or operation? Is the detail across the various blocks consistent? Probing along similar lines will also test the adequacy of results of the more inexpensive C efforts. Is there any significant variable or module omitted? Is there a high order term or additional detail that should have been included?

Figure 7 and the foregoing list of symptoms, in conjunction with probing along the lines indicated previously, allow an engineer or his manager to exercise some control over analytical efforts. Hopefully this attention will pay off in successful research efforts rather than just 'furnish additional insight'. At the very least, such probing should result in the consideration of more than one approach to a solution of a given problem.

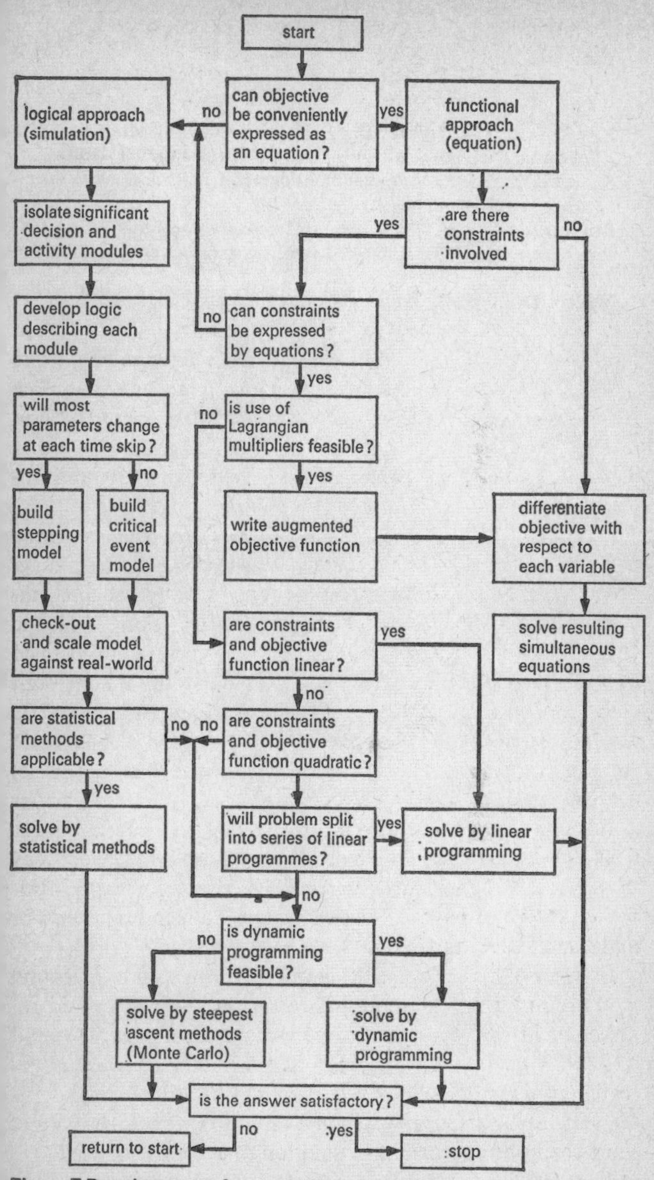

Figure 7 Development of an analytical approach

References

ACKOFF, R. L. (ed.) (1961), *Progress in Operations Research*, vol. 1, Wiley.

KAVANAGH, T. F. (1963), 'Decision structure tables – a technique for business decision making', *J. Indust. Eng.*, September-October.

SASIENI, M., YASPAN, A., and FRIEDMAN, L. (1959), *Operations Research – Methods and Problems*, Wiley.

SOKOLNIKOFF, I. S. (1939), *Advanced Calculus*, McGraw-Hill.

VAZSONYI, A. (1958), *Scientific Programming in Business and Industry*, Wiley.

WILDE, D. J. (1964), *Optimum Seeking Methods*, Prentice-Hall.

Part Two Applications of Linear Programming

What salaries should executives receive? What is the cheapest diet which, provided of course that he can bear to eat it, is sufficient to keep a man alive and well? These are just two widely different problems which can be solved by linear programming and between these extremes lies a whole spectrum of applications in agriculture, industry, economic planning and many other fields.
The first paper in this part illustrates linear programming in action in the form of a case study in forest management. The next two papers give applications to production planning in industry: De Garmo's to foundry production and Smith's to the production of transistors. The final paper shows an attempt at the economic planning of agricultural production in the United States.
While these papers are quite representative, they obviously cannot cover anything like the whole range of possible applications. Neither is it within the scope of this book to present in detail the basic theory of linear programming. References to further material on the standard techniques and applications of linear programming are provided in the list of additional references at the end of the book.

2 P. A. Wardle

Forest Management and Operational Research: A Linear Programming Study

P.A. Wardle, 'Forest management and operational research: a linear programming study', *Management Science*, vol. 11, 1965, pp. B 260–70.

Introduction

A great deal of the difficulty of the job of Management stems from the complexity of the problems with which the manager is confronted. Even if information about the quantities he has to control is available, it is frequently so intricate in detail and in the inter-relationships involved as to make it impossible to use in its complexity. His skill has therefore consisted in making the most promising approximation. He has been helped by economists who have directed his attention to the quantities most correctly representing his success and by work study and industrial engineers who have evaluated the physical possibilities. The intricacy of the problem remains.

As a result of the pioneer work of mathematicians and economists in the Thirties and the urgency of organizational problems in the Second World War, both in the development and production of equipment and the organization of the operations themselves, mathematical systems have been developed which allow such problems to be represented in their complexity. The development of computers has made their solution a practical proposition. Indeed, these developments have allowed the growth to the status of a discipline in its own right of Operational Research – the application of mathematical techniques to problems of organization with the objective of optimizing the performance of the system.

In this note the technique known as linear programming is applied to a problem of allocation which involves interaction between particular choices. Many management problems are further complicated by being dynamic – that is involving interaction through time as well as between variables – or they involve uncertainty. Such problems also lend themselves to numerical analysis. Through the description of an example it is hoped to demonstrate that the information available to management is both improved and extended by such analysis.

The Problem of Interaction

Before going on to the main problem it is desirable to be quite clear about the difficulty we are attempting to overcome in the allocation problem.

If the situation were such that one could buy discrete units of resources of exactly the right volume to carry through the particular project and the only restriction on the projects to be done was the amount of money available to buy resources, then by assessing the profitability of all possible projects and picking projects starting from the most profitable and working downwards until the available money is used up, the best collection of projects will be obtained. In practice the choices confronting the manager are not so straightforward. In addition to the amount of money available being limited, the raw materials at his disposal are restricted and a large part of the cost of operation may already be committed in capital equipment. Though the financial objective of the enterprise may be paramount, pursuit of this objective may be restricted by other requirements imposed from outside. In these circumstances, although the relative profitability of individual projects can be evaluated, these make varying

Table 1
Cost and Net Return of Alternatives

Operation	*Cost £ per acre*	*Net return £ per acre*
1. Fell hardwood and plant pine	50	250
2. Fell hardwood and regenerate	10	220
3. Retain hardwood (postponing action to a later period)	0	180

demands on the raw material and existing capital equipment and fulfil to varying extents the other requirements of the enterprise. The manager has not only to decide between straightforward alternatives but to pick the best from a collection of alternatives, each of which interacts with the remainder.

The following example illustrates in very simple terms the nature of the problem. Suppose a forester has 100 acres of forest with over-mature hardwood timber worth £200 per acre if felled in the coming period; in Table 1 the operations open to him are listed with their cost in immediate resources and the net return which results per acre. The net returns consist of the value of timber felled plus the net value expected from the replacement crop, all discounted to the present.

The objective of the enterprise is to get the largest total net return. Pursuit of this objective is subject to the constraint that only £2500 is available to meet costs whatever programme is chosen. Table 2 gives examples of possible alternative courses open to the enterprise and the total net return resulting. The obvious course of action is to do as much as possible for the most profitable alternative, fell hardwood and plant pine. In the absence of any budgetary constraint this would lead to the most profitable result. Because of the restricted budget it is in fact possible to treat only half the area in this way and one is forced to adopt the least profitable alternative, to retain hardwood, on the remainder of the area. One can do better by adopting the second alternative, fell hardwood and regenerate. The best course of action is to adopt a combination of treatments so that pine is planted to an extent that leaves just sufficient funds to allow the regeneration of the remainder. In this example involving only three possible operations, one area and one restriction, the problem of deciding the best combination is easily handled. It is immediately apparent however that if a number of areas were involved each with a set of possible operations making varying demands on resources, and of varying value, and other restrictions were imposed on pursuit of the objective beside the budget, then the problem may no longer be easily solved by trial and error as shown in Table 2. Determination of the optimum felling programme for the New Forest involves this sort of complexity. The analysis described below was part of the analysis which formed the basis on which the felling programme in the Working Plan was decided.

Table 2
Courses of Action

Operation	*Result Total net return £*
1. Fell hardwood	
and plant pine on 50 acres	12,500
retain hardwood on remaining 50 acres	9000
(half the area is planted, all resources used up)	
Total	*21,500*
2. Fell hardwood	
and regenerate hardwood on 100 acres	
Total	*22,000*
(whole area restocked, £1,500 left over)	
3. Retain hardwood — *Total*	*18,000*
(no planting done, £2,500 left over)	
4. Best combination	
Fell hardwood over whole 100 acres	
and plant 37½ acres to pine	9375
regenerate 62½ acres to hardwood	13,750
(whole area restocked and resources used up)	
Total	*23,125*

Felling Programme for the New Forest

The prime objective in the management of the New Forest is a financial one, in keeping with the policy of the Forestry Commission. Pursuit of the financial objective is subject to certain well-known restrictions which derive from the legal status of the

New Forest and from its importance as a holiday area. Over a large part of its area – the Open Forest – commercial forestry is not permitted, and over a large part of the woodlands – the Ancient and Ornamental woodlands – forestry is restricted to the maintenance of the woods as a visual amenity. Areas where commercial forestry is not permitted have been excluded from the problem. Areas where the costs of postponing action are comparatively trivial, e.g. larch crops which might ideally be treated in the period, but where postponement will only cost a few pounds, have also been excluded. In fact the problem whose solution it was decided would be of most immediate value in planning the management of the New Forest was that of choosing the optimum felling programme in hardwood, conifer, and mixed high forest within the statutory inclosures, the felling programme to be for the next ten year period. The data about crops and site types was obtained from the Working Plan survey, while the restrictions on management were decided by the local staff to conform with amenity requirements and the capacity of the labour force. The criterion for financial success was taken to be maximum net discounted revenue, and in formulating the problem the addition to cost and revenues expected to result from alternative courses of action was discounted to the present.

Definitions

The problem can be formulated if the following are defined:

(a) The objective.

(b) The restrictions or constraints on the pursuit of the objective.

(c) The various discrete alternatives open to management – these can usefully be called 'activities.'

(d) The relationship between each activity and particular constraint and,

(e) the value associated with a unit of each activity.

Taking these in turn:

(a) The objective is taken to be to maximize net discounted revenue, subject to

(b) the constraints – these in physical terms are:

(i) The volume felled in the period from conifer and mixed high forest not to exceed 4·16 million h. ft.

(ii) Volume felled in the period from old hardwood not to exceed 2·44 million h. ft.
(iii) Area treated (as distinct from treatment being postponed) not to exceed 5000 acres.
(iv) The conifer area resulting from the chosen programme not to exceed 3845 acres (this is to ensure that the area of conifers does not exceed 50 per cent of the area of the whole forest).
(v) 500 acres to be planted with hardwood/conifer mixtures.

The total area considered in this problem is some 8500 acres out of the total Working Plan area of some 30,000 acres.

(c) *Activities* (see Table 3). There are in the potential felling area a number of crop classes which may be treated in a variety of different ways.

The crop classes are: old hardwood, old conifers, mixed high forest, and bareland. The old hardwoods are further broken down into classes with varying volumes and quality of timber per acre, some with a viable understorey and some without. The alternative treatments are felling, followed by planting either of a conifer or a hardwood/conifer mixture; where an understorey exists felling may be followed by retention of the understorey or enrichment as well as the planting alternatives. Any treatment may be postponed to the next period. The treatment of a particular crop class in a particular way constitutes an 'activity'.

(d) *Activities and Constraints* (see (c) above and Table 3). The relationship between the activity and a particular constraint depends on the nature of the constraint. An example illustrates the relationship: felling and planting with conifer one acre of conifer high forest yields 4000 h. ft., thus the relationship to the conifer volume constraint (i) is 4000. It yields no broadleaved volume so the relationship to the broadleaved volume constraint (ii) is 0. One acre of felling and planting is one acre treated, so the relationship to the treatment constraint (iii) is 1. Similarly for the conifer area constraint (iv) the relationship is 1. Felling and planting one acre of conifer high forest means one acre of that class is treated but none of any other class so that the relationship to the constraint on the area treated (v) in the conifer high forest class is 1 but the relationship to constraints on area treated in all other classes is 0.

Crop Types and their Treatment Area : Value and Volume

Crop Type	*Old hardwoods*										*Conifer and mixed*		*Bareland*
	Complete understorey A1	*No understorey A2*	*Partial understorey A^{u}2*	*Complete understorey B1*	*No understorey B2*	*Partial understorey B^{u}2*	*Complete understorey C1*	*No understorey C2*	*Partial understorey C^{u}2*	*Complete understorey D1*	*Conifer high forest CHF*	*Mixed high forest MHF*	*Bareland*
Area of crop types (acres)	12	1897	500	197	523	130	39	646	170	345	1598	405	1761
Treatment Value *NDR £ per acre*													
1. Fell and plant conifer	287	287	287	207	207	207	157	157	157	287	487	337	87
2. Fell and plant hardwood/ conifer mixture	215	215	215	135	135	135	85	85	85	215	415	265	15
3. Fell and retain understorey	228	—	—	148	—	—	98	—	—	228	—	—	—
4. Fell and enrich understorey	—	—	292	—	—	212	—	—	162	—	—	—	—
5. Postpone treatment 10 years	204	204	204	148	148	148	112	112	112	204	371	264	61
Volume if felled h.ft. p.a.	2000	2000	2000	1200	1200	1200	700	700	700	2000	4000	2500	—

(e) *Value*. The value of each activity is the net discounted revenue associated with one acre of that activity. In calculating the NDR the value of the existing crop and the NDR of the replacement crop are taken into account: felling one acre of conifer high forest yields on average £400, the net discounted revenue expected from the conifer replacement crop is £87 so that the net discounted revenue associated with the activity, felling conifer high forest and replanting with conifer is £487 per acre. The NDR where treatment is postponed is the highest NDR expected to be possible in the next period discounted over ten years. The relative sizes of these values are important in determining the optimum course of action.

Formulation

The problem may now be set up as a linear programming problem which may be represented in the following manner:

Objective function: maximize $Z = cx$.

Subject to a set of constraints

$$Ax = b,$$
$$x \geqslant 0,$$

where Z is the total net discounted revenue.

Where x represents the set of activities

c the value coefficients

b the constraints and

A the matrix of coefficients relating the activities and the constraints.

The system consists of a set of simultaneous linear equations in which the number of variables exceeds the number of equations. The number of feasible solutions to such a set of equations is large, the system required, however, that that solution is selected which results in maximum net discounted revenue.

The statement of objective and constraints and the detail presented in Table 3 are sufficient to set up the linear programming problem which consists of eighteen equations involving about seventy variables, to be solved simultaneously. The process of computation is iterative. First a basic feasible solution is found

Table 4
Summary of Result

	Crop type	Treatment	Area (acres)
Old hardwoods:			
high volume and complete understorey	A_1	Postpone	12
high volume—no understorey	A_2	Plant hardwood (a)	20
		Postpone	1877
		Total	*1897*
high volume—partial understorey	Au_2	Enrich	500
intermediate volume—complete understorey	B_1	Postpone	197
intermediate volume—no understorey	B_2	Postpone	523
intermediate volume—partial understorey	Bu_2	Enrich	140
low volume—complete understorey	C_1	Plant hardwood	39
low volume—no understorey	C_2	Plant conifer	582
		Postpone	64
		Total	*646*
low volume—partial understorey	Cu_2	Enrich	170
As A: but younger	D_1	Plant hardwood	345
conifer and mixed conifer high forest	*CHF*	Plant conifer	1040
		Postpone	558
		Total	*1598*
Mixed high forest	MHF	Postpone	405
Bareland		Plant conifer	1665
		Plant hardwood	96
		Total	*1761*
Total area treated in decade			*4587*
Resulting net discounted revenue			*£1,840,000*

Note (*a*) hardwood = hardwood/conifer mixture.

that is a programme of activities which is both possible and consistent with the constraints; this is then revised by substituting an activity not included in the solution (which, it can be seen, will improve the results), for some activity in the present solution. This revision of the basis requires revision of the problem which involves computation on the values of b and c and the coefficients A in the equations. This process of substitution is repeated until no further improvement in the result is possible – the optimum solution is achieved.

In the example described here this required thirty-five steps, and took about thirty minutes on the Pegasus computer.

The Solution

The solution is set out in Table 4. This consists in programmes showing the area of treatments chosen for each crop type which will produce maximum net discounted revenue together with the value of Z, the total NDR expected. This, however, does not exhaust the information available. In addition to the optimum programme, solution of the problem throws up information about the cost and benefit associated with varying the restrictions on management or carrying out different activities from those specified in the solution. In this case the operative constraints, that is those restrictions preventing an improvement in the result, are the following (see paragraph entitled Definitions (b)):

1. The requirement that the area of conifers resulting from the programme should not exceed 3845 acres (iv).

Relaxation of this constraint would result in an improvement of £25 per acre of additional conifer allowed.

2. The restriction on the volume of conifer high forest that may be felled (i).

Relaxation of this constraint would result in an improvement of £29 for an addition of 1000 h. ft. in the volume permitted to be felled.

3. The restriction on volume of old hardwood that may be felled (ii).

Relaxation here would result in an improvement of £28 per additional 1000 h. ft. permitted to be felled.

4. The requirement that at least 500 acres be planted to hardwood/conifer mixture (v).

Relaxation of this constraint would result in an improvement of £47 for each acre removed from the requirement.

5. The restriction of the area that might be treated (as distinct from being postponed) during the period to 5000 acres (iii) did not affect the solution, only 4587 acres were, in fact, treated in the solution.

This information leads one to review the formulation of constraints to ensure that they correctly represent restrictions on the management of the forest. An earlier formulation of the problem, for example, did not include any requirement to plant hardwood/conifer mixtures. The solution resulted in no hardwood planting being included; the local staff, however, believed some hardwood planting to be essential to conform with requirements of visual amenity. The present solution shows that if this requirement can be fulfilled by other means the financial result will be substantially better. Introduction of the constraint in fact led to a reduction of £24,000 in the expected NDR.

The solution presented is not the only one which will result in maximum net discounted revenue. It is clear that where some hardwood planting is to be done, if it is assumed that the net discounted revenue from hardwood planting and that from conifer planting is the same on all areas – as has been done in this problem – then it does not matter on which of the areas to be planted the 500 acres of hardwoods are put. The species planted may therefore be decided on other grounds, such as recognizable site variations not considered in the problem, or local amenity requirements. Other alterations which can be made at zero cost (that is without affecting the value of NDR achieved) involve the interchange of treatments of high volume old hardwoods A_1, A_2 or D_1, which in this formulation prove to be to all intents and purposes the same crop type. One might therefore choose to substitute treatment of A_2 for that of D_1, the younger crop type, on the supposition that the latter is less likely to come to any harm if treatment is postponed. Some other activities can be introduced at low cost: for example, revision of the programme to include some treatment of B_1 medium volume old hardwood and B_2 is done at a cost of £1 per acre; or postponement of treatment of low volume old hardwoods C_1 and C_2 is done at zero cost. One might have expected the inclusion of the retention of hardwood understorey in the solution; this, however, has been completely displaced by the requirement that hardwood/conifer mixtures should be planted and is only introduced to this formulation at rather high cost. In the absence of the hardwood planting constraint, retention of understorey would be selected rather than hardwood/conifer planting.

In the previous paragraphs the value of relaxing constraints and the cost of making particular alterations to the programme have been mentioned. The values and costs given refer to unit changes in the programme and the extent of change to which this value will apply is limited. To illustrate this one may cite the example of the requirement to plant 500 acres of hardwood/conifer mixtures in this problem. Relaxation of this constraint is worth £47·5 per acre, whereas the initial introduction of hardwood planting cost only £46 per acre. Achievement of the benefit indicated for the relaxation of a constraint may depend not on simple substitution of the new activities in the programme but on more general revision of the programme. Thus although the result indicates where a change may be desirable or acceptable, incorporating the change in a new programme may require revision of the problem and a new solution.

Variation of Assumptions

In formulating the problem it was necessary to assume that the crop types represented by their average condition were uniform and that the values associated with each treatment were at least in correct relation to one another, though in fact the crop types are quite variable and there is some uncertainty about the relative values of particular crop types and treatments. A great merit of the analysis is that once the problem has been formulated the solution indicates the quantities in the problem that are critical. If one has doubts about the correctness of particular assumptions or wishes to revise the restrictions, this is no less than part of the exercise, indeed it was this part which one hesitated to carry out in the absence of computer aids when one was confronted by a series of arduous computations every time the implications of a new assumption were to be explored.

Value of the Analysis

The question remains, given the added information and some assurance that the chosen programme is the best, is that best so much better than the programme which would have been chosen without the computation? In Table 5 a programme has been worked out by selecting first the most valuable alternatives and continuing down the list in order of value until one of the restrictions prevented one going any further. The resulting net discounted revenue is £1,823,000, that is about £20,000 or 1 per cent below the optimum net discounted revenue of £1,840,000.

If the problem is set out as illustrated in this paper, by careful appraisal of the volume and value ratios, it is possible in this problem to select a programme which approximates the optimum programme most closely.

As has been demonstrated however, the result of this analysis is not only an optimum programme of treatments, which in itself may have been difficult to arrive at in the absence of the analysis, but also information about the effect of changes in the programme on the result. In many cases it may be that this 'sensitivity analysis' – the information about the value of change in constraint or introducing new activities – may be the most valuable information resulting. It is this information which provides critical guidance on the direction management should take.

Summary and Conclusions

Many management decisions do not involve clear cut choices between simple alternatives but rather the reconciliation of alternatives which conflict with one another and are variously affected by restrictions on management. The best course of action in these circumstances is not immediately apparent.

A clear specification of the management problem which defines: (a) the objective to be pursued, (b) restrictions on its pursuit, (c) the discrete alternatives open to management, (d) the relationship between these alternatives and the restrictions, and (e) the contribution made by each alternative to the objective; may be sufficient to allow the best solution to be recognized.

Table 5

Summary of Result Selecting the Most Profitable Activities from Table 3 (Activities Written in Order of Selection)

Crop type	*Treatment*	*NDR £ per acre*	*Volume million h. ft.*	*Area (acres)*	*Total NDR/ £.000*	*Remarks*
Coniferous and Mixed						
1. *CHF*	Fell and plant conifer	487	4·16	1040	507	4·16 exhausted conifer volume constraint therefore treatment of remaining conifer and mixed crops must be postponed.
CHF	Postpone treatment	371	—	558	207	
MHF	Postpone treatment	264	—	405	107	
Old Hardwoods						
2. Au_2	Fell and enrich understorey	292	1·00	500	146	
3. A_1	Fell and plant conifer	287	0·02	12	3	
4. A_2	Fell and plant conifer	287	1·42	710	204	2·44 exhausts hardwood volume constraint therefore treatment of remaining hardwood crops must be postponed.
A_2	Postpone treatment	204	—	1187	242	
B	Postpone treatment	148	—	850	126	
C	Postpone treatment	112	—	845	95	
D	Postpone treatment	204	—	345	70	
5. Bareland	Plant conifer	87		1261	101	
	Plant hardwood/ conifer mixture	15		500	7	This is required by the conifer/hardwood mixture constraint.
	Total area treated in decade					4023 acres
	Resulting net discounted revenue					£1,823,000

For those cases which are too complex to be solved by inspection or simple computation, the technique of linear programming or some other form of numerical analysis, may allow solution.

In addition to the optimum solution to the problem, the use of linear programming gives information about variation of the optimum programme and changes in the restrictions on the programme, which may provide critical guidance on the direction which management should take.

3 E. P. De Garmo

Applying a Modern Industrial Engineering Technique to Two Old Foundry Problems

E. P. De Garmo. 'Applying a modern industrial engineering technique to two old foundry problems', *Journal of Industrial Engineering*, vol. 17, 1966, pp. 449–53.

For many years the Industrial Engineer has been asked the question, 'What can Industrial Engineering do for small businesses?' In more recent years, with the advent of various techniques such as applied probability and statistics, linear programming, operations research, and so forth, the question has been changed to, 'Can modern Industrial Engineering techniques be helpful to a small business?' Concurrently, some Industrial Engineers who are employed in small businesses ask, 'Need the modern techniques be part of the IE's kit of tools?' The cases on which this article is based provide a resounding 'Yes'! to both questions. In addition they illustrate how these techniques frequently can provide answers to long standing problems which previously could not be solved in any reasonably economic manner. Admittedly, two cases do not establish an inviolable rule, but the nature of them is such that they should add considerable credibility to the thesis that the modern techniques can be helpful when applied with realism and judgment.

One of the major differences that often may be encountered in applying some of the advanced techniques in a small business, compared with a large organization, is that desired quantitative data are not readily available. This can be quite a shock to the uninitiated who is accustomed to having ample quantities of data available for the asking. It sometimes seems impossible that small businesses could operate for decades – and make a profit – with so few operating data available, but many do. This lack of data can be quite frustrating when one first tries to apply a technique such as linear programming to such a situation. This was not the way the problems were presented in the book.

Another difference, and frustration, that may be encountered in a small business is the fact that, due to operating conditions, or in order to get the results accepted and used or to get the co-operation of some key member of the organization, more limiting restrictions may have to be imposed. Sometimes it is better and even more economic to make some compromises and have a fairly good solution installed and working than never to get the perfect solution installed. As will be discussed later, some of these problems were encountered in these cases, yet quite good and workable solutions were obtained.

The Major Problem

A very old problem in all gray iron foundries has been that of determining the proper and most economic mix of raw materials to charge into a cupola or electric induction furnace in order to produce castings which have the desired properties. This, of course, is the well known 'blending' problem which has been dealt with so frequently in the linear programming literature. Linear programming has been applied very successfully to this problem as it exists in large iron foundries, particularly the 'captive' foundries of the automobile industries (Metzger and Schevarzbek, 1961). In these foundries, large tonnages of castings of essentially a single chemical composition, are produced. The procedure also has been applied in moderate size alloy-steel foundries, with one such company reporting savings of several thousands of dollars per month as the result of its use. In these foundries the chemical composition of the final product is specified and closely controlled. Frequent samples are taken and analysed, and ample data are available. Further, because of the volume and constant type of production there is a fairly constant supply of 'returns' which will be discussed in detail later.

The study reported here was for the purpose of determining whether linear programming could be of value in solving the blending problem in typical, small, job shop iron foundries where the conditions often are very different from those existing in the types of larger foundries cited previously. A brief survey was made of the operational practices of several small iron foundries and two were selected for a detailed study on what could be accomplished through the application of linear programming.

An Important Secondary Problem

One of the characteristics of nearly all foundries is that not only may the charge of a melt be composed of four to seven raw materials from a list of eight to twelve available materials, each with its particular chemistry and cost, but a considerable portion of each charge consists of 'returns' – the sprues, risers and runners which have been cut from previous rough castings. In a large captive foundry the percentage of returns in each charge may be fairly constant. In the job shop foundry, on the other hand, because a widely differing variety of castings is produced from day to day with very different amounts of gates and risers, the amount of returns may vary from 10 per cent to 50 per cent of the charge. While these returns have the desired chemical composition of the finished melt, they may not be the most economic raw materials for the charge due to the fact that the melting process may, in itself, alter the chemical composition, as in the case of melting in a cupola. However, from the viewpoint of determining the least-cost mix of raw materials through the application of linear programming, the real 'sticker' is, 'What is the true value (cost) of the returns?' Obviously, with the possibility of returns constituting a large percentage of the charge, if an economic mix of raw materials is to be determined, the correct value of the returns must be known.

Since all the foundries involved in the survey and study had been in operation for a good many years and were quite successful, one naturally might assume that they had ready and satisfactory answers about the cost of their returns. Such an assumption would be erroneous. It was found that few had ready answers and none was satisfactory in the light of a little scrutiny. Two examples will suffice. In one instance, a responsible staff member of a foundry stated that the value of its returns was $110 per gross ton. On inquiring about the basis for this figure, the employee said that a couple of years previously, some customer, for an unknown reason, had purchased a couple of tons of returns at $110 per gross ton. Since then, the foundry has used this figure as the value of returns, even though they had never sold any more. In a second case, the accountant of one foundry said that the value of its returns was $242 per gross ton. His premise was that the returns had the same chemical composition as the finished castings sold by the company, therefore the returns had the same value as the finished castings, less finishing costs! Clearly, a range of $110 to $242 per ton leaves much to be desired when attempting to determine the most economical mix of raw materials, particularly when these materials may constitute up to 50 per cent of the charge.

Even though the extremes cited above might lead one to conclude that small foundries are poorly managed, such a conclusion is not necessarily justified. While it quickly became clear that none of these foundries had a satisfactory cost figure for its returns, it also was clear that several recognized that their figures did not have a sound basis. Further, a little thought made it clear why this situation existed, and why a new Industrial Engineering technique – linear programming – could provide the solution.

It is evident that the true 'cost' of returns, as a raw material, is the price at which it can be sold, or, if there is no market, the cost of the most economical mix of raw materials which will provide the same chemical elements and for which it can be substituted. Obviously, for most foundries it is the latter cost which applies, since there virtually never is any market for returns. Some of the foundries recognized this fact, but how were they to know whether they were using the most economical mix of raw materials, particularly when they virtually always were using some returns in their charges? Without linear programming this was much more difficult to know. Thus, their failure to have solid cost figures for returns is not surprising, and a study that started out to be concerned primarily with a practical blending problem in small foundries ended by also providing a simple solution to a second, but just as old and perhaps as important, problem.

Some Practices Encountered in Small Job Shop Foundries

With a clear solution in view for the costing of returns, there were a few other rocks in the stream of progress. Some of these are typical of those frequently encountered when attempting to apply modern management techniques in a small business. Often, the desired quantitative data are lacking. The following instances are typical, although perhaps somewhat magnified by certain characteristics of a small foundry. To one who is not familiar with the industry, some of the practices of small, job shop iron foundries may be quite startling. For example, a foundry's staff may not know the complete chemical composition of its product with any exactness. Further, it may not care! To add to this seemingly unbelievable situation, if the staff does know what the exact chemical composition is, it may have no sound scientific basis for justifying the composition used, beyond the fact that the customers like the product.

Now, lest the uninitiated conclude that such foundries are beyond the pale of sanity, the reasons for such operating procedures will be examined. The purchaser of a casting, particularly a gray iron casting, is interested in certain physical properties, notably strength and machinability. Consequently, such castings are produced, and bought, to meet certain physical property specifications. For example, the major output of one of the foundries studied was class forty gray cast iron. This meant that it must have a minimum tensile strength of 40,000 p.s.i. On most orders there was no other specific requirement. The foundry's customers, however, knew that they could depend on its castings having good machinability and not being hard. The foundry had built an excellent reputation for always providing castings above minimum requirements and it tried to maintain a minimum tensile strength of 45,000 p.s.i., rather than the called-for 40,000 p.s.i., and exceptional machinability. Consequently, over the years it had added certain alloying elements to its melt which it felt would assure this extra quality even though adequate properties might have been obtained with lesser amounts of alloys, particularly if tighter melting and pouring controls were maintained.

The difficulty with this situation is that the relationships between these desirable physical properties and the chemical composition of the melt are not precise. Those who are familiar with the property know that machinability cannot be defined precisely. It means different things to different people who have different machining objectives and criteria. Those who have gone deeper into the subject know that machinability can be improved by several chemistry changes, but there is not complete agreement as to the precise relationships. Further, there is not complete agreement as to how the chemistry affects some of the other physical properties of cast iron. Also, since some of the properties are affected by several elements, the total result may be relationships which are nonlinear. A good example is that of tensile strength. A study of the literature revealed that tensile strength is a function of an alloy factor based on the effects of molybdenum, chromium and nickel, as shown in Figure 1. Consequently, in order to apply linear programming, it was necessary to approximate the curve by two straight lines, as indicated in Figure 1.

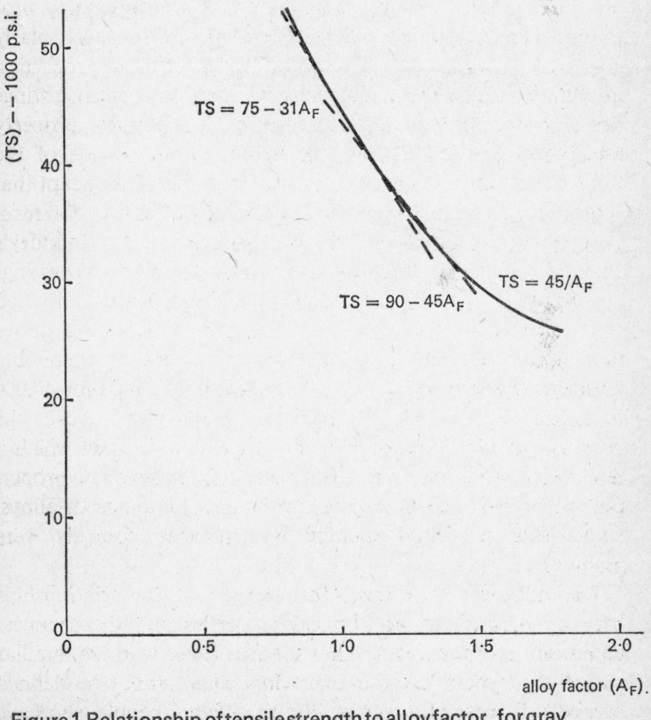

Figure 1 Relationship of tensile strength to alloy factor, for gray cast iron

Another example of the type of practical compromise that may have to be made when dealing with small businesses was that, although there was no scientific evidence to indicate that the most economic amount of one material could not be used successfully in the cupola, the melting foreman did not agree with the theory and insisted that the maximum that could be used was only about 60 per cent of the theoretical optimal value. He had a memory of previous trouble with the charge hanging up in the cupola lining when he had used a similar material several years before. Management did not wish to arouse the non-cooperation of the only man they had who was capable of operating the cupola; enter one more restrictive equation! In applying new techniques in a small business the personnel factor may be much more important – and restrictive – than in a large organization.

In another case it could be shown that the most economical mix for an induction furnace called for obtaining the required carbon from carbon briquets. However, it also was essential that the foundry obtain a certain daily output during day-shift operation, and this required a melting cycle which was too short to permit the carbon, in briquet form, to be completely dissolved in the iron. Consequently, a restriction in the amount of carbon briquets had to be included.

Results

After all the various restrictions – some technical, others human – were determined for the case of the foundry that worked to physical properties, it was decided that the bounded items and requirements in Table 1 would have to be dealt with.

Once it had been determined that these factors and requirements would provide a product with the desired properties, obtaining the solution by linear programming was straightforward. Table 2 shows the results with the computed mixes compared with the mixes actually being used by the foundry.

The first matter of importance shown in Table 2 is the cost of the mix when no returns are used. Obviously this figure of $97.50 per gross ton represents the true cost of returns since this is the minimum cost for which a product could be obtained that would be interchangeable with the returns. Further mention will be made concerning this later. It will be noted that the least-cost mix represents a saving of $3.80 per ton as compared with the cost of the mix used by the foundry when no returns were used.

With the true value of the returns known, it was possible to determine the least-cost mixes using various percentages of returns and to compare these with the costs of the mixes then being used. These values also are shown in Table 2 for 10 per cent, 30 per cent and 50 per cent of returns. It will be noted that the potential savings through using least-cost mixes ran as high as $6.20 per ton. The average saving, taking into account that the average use of returns was about 20 per cent, was approximately $5.50 per gross ton. Since the foundry was producing about 6000 tons per year, the potential annual saving was roughly $33,000.

Table 1

Item bounded	*Requirement*
Total Carbon	3·20–3·50%
Silicon	1·80–2·30%
Manganese	0·50–0·80%
Phosphorus	0·1–0·3%
Sulfur	0·07–0·12%
Nickel	0·9–1·2%
Chromium	0·2–0·35%
Molybdenum	0·5–0·7%
Tensile Strength	45,000 p.s.i. (minimum)

Another potential saving of an undetermined amount was connected with the fact that the analysis revealed that two of the raw materials – car wheels and structural steel scrap – were

Table 2

Comparative Costs and Resulting Saving per Gross Ton Using Least-Cost Mix and Standard Foundry Mix.

	No returns		*10% returns*		*30% returns*		*50% returns*	
	LCM	*FM*	*LCM*	*FM*	*LCM*	*FM*	*LCM*	*FM*
Cost (per gross ton)	*$97.50*	*$101.30*	*$99.00*	*$105.20*	*$97.70*	*$102.00*	*$93.60*	*$94.20*
Saving	*$3.80*		*$6.20*		*$4.30*		*$0.60*	
Materials (pounds per 1000 pound charge).								
Pig iron	—	200	—	200	—	200	—	200
Car wheels	—	—	—	200	—	—	—	—
Scrap iron	910	600	850	300	640	300	430	—
Steel rails	90	200	50	200	60	200	70	300
Structural steel	—	—	—	—	—	—	—	—
Returns	—	—	100	100	300	300	500	500
Ni*	10·5	10·5	8·75	8·75	7	7	3·5	3·5
Cr*	5	5	5	5	2·5	2·5	2·5	2·5
Si*	2·5	5	—	12·5	—	5	—	7·5
Mo*	10	10	10	10	7·5	7·5	5	5
Composition (%)								
Carbon	3·4	3·3	3·47	3·22	3·44	3·3	3·42	3·21
Silicon	1·95	1·86	1·8	1·93	1·8	1·91	1·8	1·92
Manganese	0·76	0·75	0·75	0·67	0·73	0·72	0·72	0·69
Phosphorus	0·27	0·2	0·29	0·1	0·29	0·1	0·28	0·1
Sulfur	0·11	0·1	0·11	0·09	0·11	0·1	0·11	0·07
Chromium	0·33	0·34	0·35	0·35	0·21	0·24	0·21	0·3
Nickel	1·0	1·0	0·95	0·94	1·02	1·01	0·9	0·9
Molybdenum	0·5	0·5	0·58	0·58	0·6	0·6	0·62	0·62

* Added to ladle. *LCM*=least-cost mix. *FM*=standard foundry mix.

not economical to use and so no longer needed to be purchased, stored and handled separately in the scrap yard.

Comparison of the composition data shows that the results obtainable by using the least-cost mixes would be almost the same as those currently resulting from more costly mixes. Since every restriction which the foundry had specified in conformity to what it believed would be workable operating practice had been built into the program, there seemed to be no reason to doubt the results. Further, since all restrictions requested by the foundry had been included, it was easy to accept the results.

The importance of having a sound cost value for the returns available cannot be overemphasized. Even if the least-cost mixes had not been available, a correct cost figure for returns would have enabled the foundry to compute valid costs for the mixes it actually was using and, thus, to operate more intelligently and probably with greater economy, even if not with maximum economy. Without a valid cost figure for returns, the cost figures for all the mixes which included returns undoubtedly would be wrong. Consequently, a decision based on two or more incorrect costs could hardly be expected to be a sound one.

A similar procedure provided a cost figure for the returns of the foundry that used induction furnaces and also gave the desired information concerning the proper mix of raw materials to provide minimum cost. However, there were several additional interesting and valuable results in this case. This foundry produced two grades of gray cast iron, which will be called A and B, and ductile iron, C. The costs of the A and B returns were approximately the same, but those of C were about $15 per ton greater. An important question was whether the three types of returns had to be segregated and handled and used separately. The study showed that using B returns in iron C would reduce the cost of C by about $3 per ton. However, the consequent reduction of B returns for use in B iron would increase the cost of that iron by $1.75 per ton. Since the production of B was over twice that of C, it was not economical to use B returns in C iron.

Previously it was mentioned that the amount of lump carbon used in the induction furnace charges had to be limited due to melting cycle restrictions. With this restriction, the cost of the returns per ton was $2.25 higher than could be obtained without the restriction. Once it was known that a less costly mix could be obtained by using a longer melting cycle, arrangements were made to operate the induction furnace between 5 a.m. and 8 a.m., using the longer cycle and most economical mix and putting the output into a holding furnace from which it would be poured as soon as the regular day shift arrived. Thus, greater over-all economy was possible because the application of linear programming not only gave the economic solution with the melting cycle restriction, but also revealed that there was an added possibility.

A question still remains: 'What would it cost a small foundry to make use of linear programming?' The total computer time amounted to 0·12 hours. At $230 per hour this was a cost of $27.60. With stock programs available which readily can be adapted to such work a competent Industrial Engineer should not require more than a week to obtain the required data and get an initial solution. Thus, about $225 would be involved for the initial study and solution. This is a small investment for an annual saving of over $30,000, which was the potential in one of the cases. To keep the operation updated as prices change would be very inexpensive. Thus, it is clear that the use of modern Industrial Engineering techniques definitely can be of great benefit to small businesses, if they are applied with good judgment. Further, Industrial Engineers in small businesses can not afford to feel that such techniques are applicable only to corporate giants.

Reference

METZGER, R. W., and SCHEVARZBEK, R., (1961), 'A linear programming application to cupola charging', *J. Indust. Eng.*, vol. 12, no. 2.

4 S. B. Smith

Planning Transistor Production by Linear Programming

S. B. Smith, 'Planning transistor production by linear programming', *Operations Research*, vol. 13, 1965, pp. 132–9.

The difficulties involved in planning and controlling production of transistors stem from three sources. First, there is little product standardization in the industry and most sales are to individual customer specifications. The variety and changing character of customers' needs present a complex, moving target for production. Second, the technology is still in its early stages, and much remains to be learned about how the choice of materials specifications and processing variables affects the distributions of electrical parameters in the final product. Finally, there is considerable stochastic variation in the product resulting from the use of components of microscopic size and raw materials that require a high degree of purity and are extremely susceptible to contamination. Because of these factors, a substantial part of the total manufacturing cost is incurred by rejection of work in process and mismatch between the electrical characteristics of the output from production and sales requirements.

The purpose of this paper will be to describe a linear programming model of transistor production that was developed to deal with these problems. The model was used to select materials, processing variables, and activity levels, with the objective of satisfying sales and inventory requirements at minimum over-all cost. This model was developed in 1958 in the Semiconductor Division, Raytheon Company, and was subsequently used for planning and controlling production of several families of germanium alloy transistors.

The Production Process

A typical production process for germanium alloy transistors starts with germanium dioxide. This white powder is reduced in a hydrogen atmosphere to obtain the germanium element, which is then melted and cooled to form solid bars of metallic germanium. The bars are purified by zone refining; one end of the bar is heated to form a molten zone that is moved slowly along the length of the bar. Most impurities are swept along in the molten zone to be concentrated at one end of the bar which is then chopped off. Purity is measured by resistivity since impurities serve as current carriers.

The purified germanium is melted in a crucible. A germanium seed is lowered into the melt and slowly withdrawn. The molten germanium adheres to the seed and forms a single crystal as it becomes solid. Controlled amounts of impurity, such as antimony or gallium, are added to the melt to obtain the desired crystal resistivity.

The crystal is sliced with diamond saws and the slices are ground. They are then scribed and broken into tiny square wafers. These are etched to reduce their thickness to specification and remove surface impurities. Final measurements may be about 0·05 × 0·05 × 0·03 inches.

The wafers are loaded in cavities in graphite jigs. Two indium spheres, a collector and emitter, are also inserted in the cavities, one on each side of the wafer. The jig is then passed through a furnace where the germanium wafer melts and alloys with the two spheres. Upon cooling, a sandwich-like subassembly is produced, which is the heart of the transistor.

The geometry of the two junctions in the subassembly, including alignment, area, parallelism, regularity, and distance apart, has a critical effect on the operating characteristics of the finished transistor. The subassemblies are tested electrically, and if penetration of the indium has been too deep and the fusion areas meet, the subassembly shorts and is rejected. If the test indicates that fusion has not been deep enough, the subassembly may be sent through the furnace again.

The subassembly is mounted on a stem that consists of a base and three leads. The leads are connected to the collector, the emitter, and the wafer by welding, soldering, and bonding. The mount assembly is etched to remove any surface material damaged by the mechanical operations. The subassembly is then encapsulated by welding a metal can to the stem; this provides protection against moisture, shock, and vibration. Finally the transistor is plated and marked.

In final test, units are extracted from a transistor family to satisfy the specifications of particular product types. The term transistor family means transistors made with the same components, physical configuration, and target specifications. Product type refers to the way in which the transistors are defined when they are sold; this may be by industry standard, company standard, or customer specification. All units from a family are first tested automatically for a number of the more important parameters, and are sorted into an interim inventory consisting of up to 2000 categories. From there they are withdrawn and tested further against the specifications for product types. Tests on sixty different combinations of parameters and test conditions against the specifications for one hundred different product types may be performed during the course of a month.

Formulating a Model

From an economic standpoint, the decision as to which subassemblies to mount, which to refire and which to scrap is of critical importance. Subassemblies are placed in categories based on the results of electrical testing and there is a strong correlation between the category of subassemblies mounted and the yields of finished transistors to product types at final test. Furthermore, the cost of the subassembly is small relative to the material and labor costs incurred in subsequent operations, and a substantial saving in assembly and testing costs can be achieved by selecting subassemblies to be mounted in such a way as to minimize the mismatch between production and demand. On the other hand, being too selective in choosing subassemblies could reduce subassembly yield to a point when the increased cost of subassemblies would outweigh the savings in assembly and final test.

Based on these observations, a linear programming model of the feed-mix type was considered that would include various categories of subassemblies as raw materials. This formulation introduced a problem of costing of subassemblies. The furnace produces a distribution of subassemblies to categories. The total cost of subassemblies depends upon the yield to production or the match between the distribution from the furnace and the distribution required for mounting obtained from the linear programming solution. However, once the requirements for mounting are known, the wafer specifications or the furnace controls would be changed to improve the match between the subassemblies produced and those required for mounting. In this way, the yield and hence the cost of subassemblies would be changed rendering the linear programming solution invalid.

This problem was solved by moving the initial cut-point for the model back one stage. Firing patterns were defined as particular combinations of wafer resistivity and furnace temperature, and the distribution of subassemblies to categories for each firing pattern was determined. The firing by pattern was included as a set of activities in the model.

The interim inventory of transistors also presented problems. Initially it was planned to consider production into each category of interim inventory and solve for the level of testing out of each category to produce types. However, this would require a model restriction on each interim category that withdrawals had to be less than or equal to starting inventory plus production into the category and would require up to 2000 restrictions. At the time, we were using the Orchard-Hays revised simplex program on an I.B.M. 704 computer, and the maximum number of restrictions allowed by this program was 255. Furthermore, even if we wrote a program to handle 2000 restrictions, determining test yields for each of the interim categories to all product types, and maintaining this data in the face of constantly changing customer requirements seemed infeasible.

It was decided that an improvement in the characteristics of the product coming into testing was much more important than improvement in the efficiency of testing itself. Therefore, realism in testing was sacrificed and the model was simplified by the use of an idealized interim inventory of greatly reduced dimensions. Although there were up to 100 product types per family in a given month, as is usually the case a small number, perhaps eight, of these accounted for most of the sales. It was estimated that if production plans were based on satisfying the requirements for these few large volume types, the requirements for the other smaller volume types could normally be met from fallout. Transistor categories in the idealized interim inventory were defined as the Venn sets of specifications for large volume product types. For example, if there were three product types, A, B, and C, there could be up to seven transistor categories, that is transistors passing the specifications for and only for ABC, AB, AC, BC, A, B, and C. Some of the specifications for types were mutually exclusive, further limiting the number of categories.

The Model

The model was designed to determine a monthly production plan that would satisfy the net demand for each product type at minimum over-all manufacturing cost.

Symbols used in the model are as follows:

1. Input data:

(a) *Demand*

D_g = demand for the gth product type in units: sales forecast plus or minus planned change in inventory.

(b) *Yields*

u_{hi} = first fire yield using the hth firing pattern to the ith subassembly category.

v_{ij} = refire yield from the ith subassembly category to the jth subassembly category.

w_{jk} = yield of mounting from the jth subassembly category to the kth transistor category.

(c) *Costs*

C_f = variable cost of first firing one subassembly.

C_r = variable cost of refiring one subassembly.

C_m = variable cost of main assembly and testing per subassembly mounted.

Z = total variable cost for the period.

(d) *Capacities*

K_f = firing capacity in subassemblies input.

K_m = assembly capacity in subassemblies mounted.

2. Variables to be evaluated:
   - F_h = quantity of subassemblies first fired using the hth firing pattern.
   - R_i = quantity of subassemblies refired from the ith category.
   - M_i = quantity of subassemblies mounted from the ith category.
   - W_{kg} = allocation from the kth transistor category to satisfy demand for the gth product type.

The objective function to be minimized is the sum of cost of firing, refiring, and assembly.

$$Z = C_f \sum_h F_h + C_r \sum_i R_i + C_m \sum_i M_i. \qquad 1$$

For each subassembly category there is a restriction that input from firing and refiring must be greater than or equal to withdrawals for mounting and refiring. Starting inventories are neglected because it was found that in critical subassembly categories they were normally insignificant.

$$\sum_h u_{hi} F_h - \sum_j v_{ij} R_i \geqslant M_i + R_i. \qquad 2$$

There is a similar restriction for each transistor category. Production into the category must be greater than or equal to withdrawals to product types. Again starting inventories are neglected, in this case because they are already deducted from gross demand to obtain net demand by product type.

$$\sum_i w_{jk} M_i \geqslant \sum_g W_{kg}. \qquad 3$$

A capacity restriction on firing states that the number of units fired plus the number refired is restricted by the total firing capacity.

$$\sum_h F_h + \sum_i R_i \leqslant K_f. \qquad 4$$

Similarly the number of subassemblies mounted is restricted by the total assembly capacity.

$$\sum_i M_i \leqslant K_m. \qquad 5$$

The final set of restrictions requires that demand for each product type is satisfied, that is, withdrawals from transistor categories to types must equal demand.

$$\sum_{k} W_{kg} = D_g. \qquad 6$$

Data Collection

Obtaining the yield data from subassembly categories to transistor categories presented the largest single problem in implementing the use of linear programming. In dealing with semi-conductor devices, laboratory results were often not reproducible on the production line. Furthermore, the type of yield data required was not normally recorded in production. By the time a transistor reached final test, there was no way of knowing what subassembly category it came from. In addition, it was not determined in final test whether a unit would pass the specifications for each major product type, the information needed for establishing yields to the transistor categories used in the model.

The required yields from subassembly to transistor categories were obtained through a series of large-scale production experiments. A tag showing the category the subassembly came from was attached at mounting to one of the leads of each transistor in the experiment. Special procedures were established to test these units to all major product types when they arrived at final test.

In order to enable us to take new product types into account in the program prior to their actual production and to reduce the need for further production experimentation, a method for simulating experiments was developed. Transistors from an actual experiment were extensively tested by parameters and the results for each unit were recorded on a punched card together with pass or fail information for each major product type. Then when a new type was to be produced, the cards were sorted against the type's specifications and a new set of yield data was obtained.

Cost data were available but had to be reorganized in a form suitable for use in the model. For example, the cost of assembly and test per subassembly mounted, C_m, was developed in the following way.

Let $1 \ldots i \ldots n$ = operations, mounting through final test.
Y_i = yield of the ith operation in per cent.
C_i = variable cost of performing the ith operation per unit of input.

Then

$$C_m = C_1 + Y_1 C_2 + Y_1 Y_2 C_3 + \ldots + (Y_1 Y_2 \ldots Y_{n-1}) C_n$$

$$= C_1 + \sum_{i=1}^{n} \left(\prod_{j=1}^{i-1} Y_j\right) \qquad 7$$

Operation

For a number of reasons, the first application involved a linear programming model of small dimensions. We intentionally selected a product family with relatively few major product types. In addition, there had been a high degree of mismatch between past production and demand so there were severe shortages of some types and excess inventories of others. This meant that the number of types with a net demand for the month was further reduced. Finally, the restrictions on production capacity were not used as the management first wanted to determine the optimum method of manufacture regardless of the capacity that would be required.

As a result, the model for the first program contained only ten restrictions and twenty-four variables. The input was punched on eighty cards and the computer running time was 7·2 minutes. Later applications to more complex families required larger models and longer running times. However, once insight was gained into the structure of the solution, it was often found possible to produce new optimal programs simply by modifying programs produced for prior months that had similar distributions of demand.

The production plan for the month was prepared directly from the linear programming solution and contained the following:

1. A firing and refiring plan showing (*a*) how many chips to first fire by resistivity range and the furnace temperatures to use, (*b*) how many subassemblies to refire from each subassembly category, and (*c*) the expected final distribution of subassemblies by category.

2. A mounting plan showing how many subassemblies to

mount from each category. In cases where subassemblies were common to different families, the plan showed how many subassemblies to mount to each family from each subassembly category.

3. A production plan showing the number of units to be produced by transistor category.

Each month on the day the plan was produced, a meeting was held of all the production and engineering managers concerned. The plan was reviewed in detail and then immediately put into effect.

Results

The first program called for major changes in the method of operation. The furnace temperature was changed to give a different distribution of subassemblies. A wider range of subassemblies was mounted, and this resulted in a larger subassembly yield. Ninety-five thousand subassemblies, which had previously been accumulated as scrap, were mounted. At final test, yield to the most critical product types increased by 80 per cent. In other words, total customer demand could be satisfied by producing only 55 per cent as many transistors as were required before. The net result was a substantial saving in manufacturing costs.

During this period, semiconductor technology was developing at a rapid pace. The manufacturing system that we modeled was subject to continual innovation involving changes in materials, components, processes, production and testing equipment, and product designs. Frequently the innovations resulted in sudden changes in yields making production far different from that expected by the current program. This also made obsolete the yield data used in programming and required us to undertake new production experiments. Sometimes further changes rendered these experiments obsolete before they were completed.

As a result of these changes, production planning by linear programming was frequently interrupted. Nevertheless, it was continued for several years and extended to other product families. Even when linear programming was not being used directly, the information gained from the experiments and previous linear programming solutions was used in managing production.

Conclusion

It has sometimes been stated that linear programming should be applied in stable situations. In this case, it was applied in what was probably the most rapidly changing industry in America. It is true that this dynamic situation made application difficult and intermittent, and that it limited extension of the model and the uses that were made of it. Nevertheless, the project proved to be highly profitable, and this was attributable, at least in part, to the very fact that the technology was new and the managers and engineers responsible had not had a long period of time to make improvements by more traditional methods.

5 E. O. Heady and A. C. Egbert

Regional Programming of Efficient Agricultural Production Patterns

E. O. Heady and A. C. Egbert, 'Regional programming of efficient agricultural production patterns', *Econometrica*, vol. 32, 1964, pp. 374–86.

Our purpose here is to employ linear programming in an analysis of interregional competition in agriculture and in the efficient location of crop production. We are concerned with regional adjustments needed to bring agricultural production into greater balance with 'food requirements', and to cause the interregional allocation of crops to be more consistent with differential changes in technology and factor prices by regions. As an analysis of agricultural production, this study also has importance for other sectors of the 122 regions delineated. The quantitative analysis suggests the amount of land to be withdrawn from crop production and shifted to less intensive uses, such as recreation, forestry, and grazing in the various regions. This shift would necessarily require larger farms and a smaller farm population. Finally, the study allows examination of national supply prices for the commodities studied and within the particular model formulation.

The Basic Study

This study is the result of an application of linear programming to determine efficient regional production patterns for specified farm commodities. While the approach is somewhat conventional, we believe that for no previous study of this large problem has there been a parallel effort in data assembly and construction. Several man-years were invested in assembling basic data and in converting them to the restraint and resources requirement vectors which serve as the quantitative foundation. The assembling and refining of data will continue for years; models and results will continue to be extended and improved, as time and resources allow. We look upon the results which follow as progress toward a useful and detailed analysis of interregional relationships in production, resource use, and structure of agriculture. We have models under way which incorporate non-linear objective functions and regional demand functions.

We made an earlier analysis of grain allocation over the United States, an elementary one which included fewer regions and only feed grains and wheat (Egbert and Heady, 1961; Heady and Egbert, 1959); the data referred to the year 1954, as a benchmark from which to launch our larger studies. The study reported here was extended to include 122 producing regions and considers wheat to be used for either livestock feed or human food, several feed grains (corn, oats, barley and grain sorghum), cotton, and soybeans. Also, the current study is devised to allow substitution among feed categories, in order that national requirements can be met in terms of both (a) locations with greatest comparative advantage, and (b) the technical substitutability among feed categories in livestock production. Too, whereas the earlier and more elementary models pointed to a time in the past, since it was necessary to establish a benchmark in time, the present study is projected to 1965 as a basis (a) for better specifying the magnitude of surplus capacity in agriculture, and (b) for suggesting policies for alleviating surplus problems.

Basic Procedure and Coefficients

Our objectives are to determine (a) which field crops should be produced in 122 agricultural regions, (b) the total acreage required to produce the nation's food and fiber requirements, (c) the amount of land which should be shifted from crop production if the nation were to cease accumulation of surplus stocks, and (d) the location of this land. Production of major field crops (corn, wheat, oats, barley, grain, sorghums, tobacco, and cotton) has been under acreage or output control most of the time since 1933. This study includes the field crops mentioned above, with soybeans added and tobacco excepted. Soybeans serve in either competitive or complementary relationship to the other crops, and their acreage and location are affected by control programs which cause land to be shifted from these other crops.

Linear programming models are used to specify which of the 122 regions might provide the nation's requirements for wheat, feed grains, cotton, and soybeans most efficiently in 1965, the efficiency criterion being the lowest total supply cost when national requirements are met. Unit production costs were estimated for each cropping possibility within each region. Cost estimates for 1965 are based on projections of trends in technology and inputs. Consumption restraints for 1965 are based on population and *per capita* income projections and on existing knowledge of price and income elasticities of demand. Since demand elasticities are low for food (e.g., the income elasticity of expenditures on food is about 0·15, with most of this for services with food, while the elasticity approaches zero for food in physical form), it is not highly unrealistic to consider domestic food demand in terms of a discrete requirement resting largely on population. The projected bill of goods to be produced for 1965 is approximately: food wheat, 1113 million bushels; feed grain 5338 million bushels; oilmeal, 316 million hundredweight; and cotton, 16·4 million bales (500-pound gross weight). These restraints limit production nationally.

Production is limited in each region by the maximum total acreage available for growing wheat, feed grains, soybeans and cotton. The first model requires that feed grain and protein from soybean meal be used in fixed proportions for livestock rations, and that soybeans be grown in rotation. The second model allows the least-cost mix to be used depending on regional variation in production costs for feed grains and soybeans. The third model allows continuous cropping of soybeans in each region, rather than requiring this crop to be combined in rotations or specified mixes with other feed grains. (One purpose of the analysis is to determine how a simultaneous shift in soybeans and cotton acreages might affect an efficient regional distribution of wheat and feed grain production.)

Crop yields were estimated by regions. The yield trend for the period 1940–59 was used to project to 1965. This method of projection assumes that technological advancement in crop production will be at about the same rate for 1960–65 as for

1940–59. Input coefficients or production costs included those for labor, machinery, seed, chemicals, and all miscellaneous items. Estimated production costs were weighted averages of several production methods used within each region, which ranged from fully mechanized to the use of horse-drawn equipment and much hand labor; production costs for cotton also included costs of hauling cotton to the gin and ginning. Only mechanized techniques were considered for soybean production, since full mechanization of this crop is nearly universal over the nation. Estimates for 1965 levels of fertilizer use were made by fitting linear regressions to the use of nitrogen, phosphorus, and potassium individually, by states. Projected changes in fertilizer use were used to estimate yield increases.

The model used to estimate national requirements considered demand for each commodity to be a function of its own price, the price of substitutes, *per capita* income, and population. Because *per capita* income and population are exogenous variables with respect to the production or programming problem, their values could be derived independently and were obtained from other independent projections. On the other hand, commodity prices are 'somewhat' endogenous with respect to the programming problem. Hence, in order to estimate requirements for programming, it was necessary to estimate approximate supply prices. While this procedure is more or less circular, our previous experience with national programming models has provided insight as to the supply price levels which might prevail. Given the prices and demand so estimated, total requirements of each commodity were estimated on the basis of a national demand model, which is not presented here but is synthesized from the many estimates of food demand made over the last decade.

Acreage constraints or maximum acreage that could be planted to the seven crops considered were the actual acreages of the seven crops planted in 1953, as production or acreage controls were not in effect in that year and, following high postwar prices, crop acreages tended to be near their physical maximum. Hence, these acreages perhaps best represent maximums that could be planted under free market conditions or efficient allocation of crop acreages. Two objectives were kept in mind in making adjustments of acreage restraints to 1965 conditions:

(1) to establish a realistic upper bound on plantings of either wheat, feed grains, soybeans, or cotton, and (2) to establish new weights for feed grain rotations that reflect trends evident in a number of sections in the country.

The Programming Models

Three programming models, summarized in Table 1, were formulated. Each region has five activities or production alternatives: (a) food wheat, (b) feed wheat, (c) feed grain rotation, (d) feed grain-soybean rotation, and (e) cotton (plus soybeans grown out of rotation for model C). The upper acreage bound for each region is that described above. The discrete demand restraints in Table 1 allow a small fraction of national production (not included in the amounts) to be produced in the regions without numbers in Figure 1. (This small production from scattered regions is not for purposes of major crop competition and is somewhat fixed and, plus seed requirements, accounts for differences between Table 1 and the total bill of goods.)

Table 1 shows that two of the models allow for ranges in feed grain and oilmeal production and demand restraints. The matrix and restraint system for these models was constructed so that if feed grain were at its maximum level, oilmeal had to be at its minimum level, or vice versa. Substitution of feed grains and oilmeal between this limit was possible. These model details permit selection of a program to produce the nation's total feed requirements most efficiently, within the bounds specified, in the sense that farmers are allowed the least-cost ration for livestock. The objective for the models is to minimize cost, given regional price differences for the national aggregate of requirements. The use of regional price differentials, if certain conditions hold true, permits the estimation of a general price equilibrium solution. While, as explained later, we do not present a general price solution in this report, we will be able to do so after further computation, as the models were designed to allow methods of analysis, extending beyond those reported in this study.

Table 1
Summary of Unique Characteristics of Six Programming Models Used in Regional Analysis of Major Field Crops

	Model regional activities	Net demand constraint levels				Programming objective
		Food wheat mil.bu.	*Feed grain mil.bu.*	*Oilmeal mil.cwt. of SBOM eq.*	*Cotton million 500-lb. G.W. bales*	
A	Food wheat Feed wheat Feed grain rotation Feed grain-soybean rotation Cotton-cottonseed	949·2	4895·9	299·4	16·1	Minimum of total costs and price differentials
B	Food wheat Feed wheat Feed grain rotation Feed grain-soybean rotation Cotton-cottonseed	949·2	4807·5 to 4943·0	283·4 to 329·4	16·1	Minimum of total costs and price differentials
C	Food wheat Feed wheat Feed grain rotation Feed grain-soybean rotation Cotton-cottonseed Soybeans	949·2	4807·5 to 4943·0	283·4 to 329·4	16·1	Minimum of total costs and price differentials

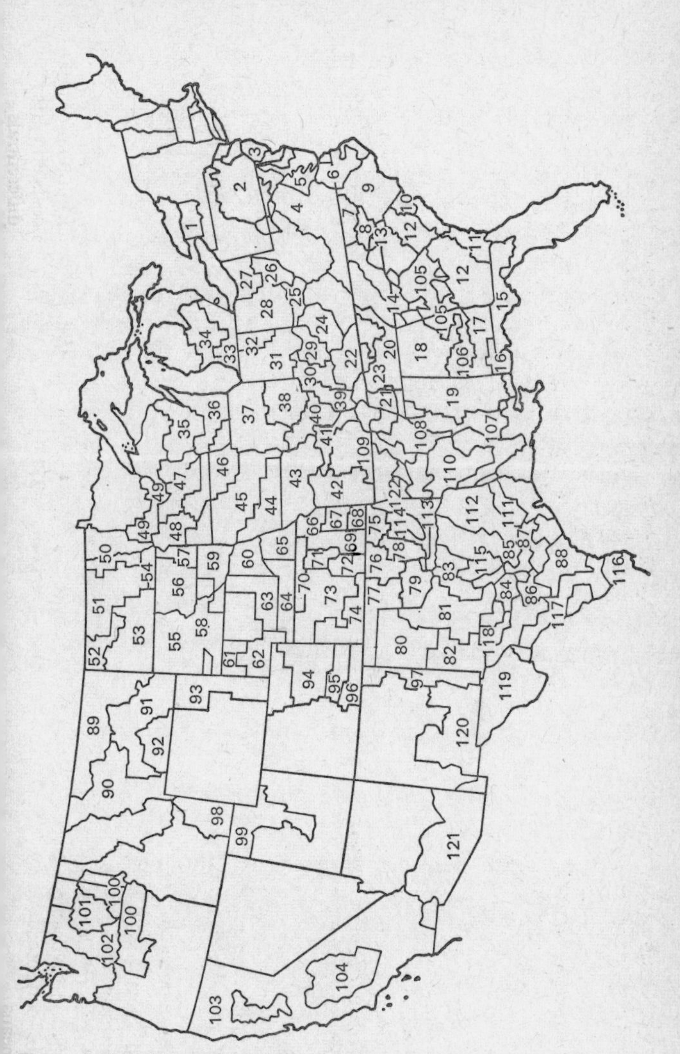

Figure 1 The 122 programming regions

Formal Models

The models now can be summarized as follows:

Let

x_{ij} be the level of the *j*th activity in the *i*th region,

c_{ij} be the cost per unit of the *j*th activity in the *i*th region,

P_k^0 be the price of the *k*th product in the base region,[1]

P_{ik} be the price of the *k*th product in the *i*th region,

d_{ik} be the price differential of the *k*th product with respect to the base region,

b'_{ij} be the primary output of the *j*th activity in the *i*th region,

b_{ij} be the secondary output of the *j*th activity in the *i*th region,

$C_{ij} = c_{ij} + b_{ij}\, d_{ik} + b'_{ij}\, d_{ik}$, where $d_{ik} = P_{ik} - P_k^0$,

s_i be the land constraint in the *i*th region,

a_{ij} be the per-unit land requirement of the *j*th activity in the *i*th region,

D_k be the national requirement of the *k*th product,

where $(i = 1, 2, 3, \ldots, 122)$,
$(j = 1, 2, 3, 4, 5$ – and 6 for model C),
and $(k = 1, 2, 3, 4)$.

The objective is

$$\min f(c) = \sum_{i=1}^{122} \sum_{j=1}^{5} x_{ij}\, C_{ij}, \qquad 1$$

which is subject to regional land constraints of the form

$$\sum_{j=1}^{5} x_{ij}\, a_{ij} \leqslant s_i, \qquad 2$$

and also subject to national requirement constraints of the form

$$\sum_i x_{i1}\, b_{i1} = D_1, \qquad 3$$

$$\sum_i \sum_j x_{ij}\, b_{ij} = D_2, \qquad 4$$

$$\sum_i \sum_j x_{ij}\, b'_{ij} = D_3 \qquad 5$$

1. Prices in base regions were used in estimating prices for other regions. (See discussion in the text.)

and

$$\sum_i x_{i4} b_{i4} = D_4, \qquad 6$$

where $k = j = 1$ is food wheat, $j = 2$ is feed wheat, $j = 3$ is feed grain, $k = j = 4$ is cotton, $j = 5$ is feed grain–soybean rotation, $j = 6$ is soybeans grown alone, $k = 2$ is feed grain, and $k = 3$ is oilmeal.

Model B is similar to model A in all respects except for the demand constraints, which are

$$\sum_i b_{i1} x_{i1} = D_1, \qquad 7$$

$$\sum_i \sum_j b_{ij} x_{ij} \leqslant D_2', \qquad 8$$

$$\sum_i \sum_j b_{ij}' x_{ij} \geqslant D_3', \qquad 9$$

$$\sum_i b_{i4} x_{i4} = D_4 \qquad 10$$

and

$$D_2' - \sum_i \sum_{j=2}^{4} b_{ij} x_{ij} = \pi \left[D_3' + \sum_i \sum_{j=4}^{5} b_{ij} x_{ij} \right], \qquad 11$$

in which D_2' is the upper bound on feed grain requirements, D_3' is the lower bound on oilmeal requirements, and π is a constant equal to the marginal rate of substitution of oilmeal for feed grains. Model C is similar to B in all respects except for the number of activities or production possibilities: One more activity, $j = 6$, is included.

Quantitative Results

The above models were used in specifying the regions which would be devoted in majority to particular crops if most efficient production patterns were used in meeting national requirements

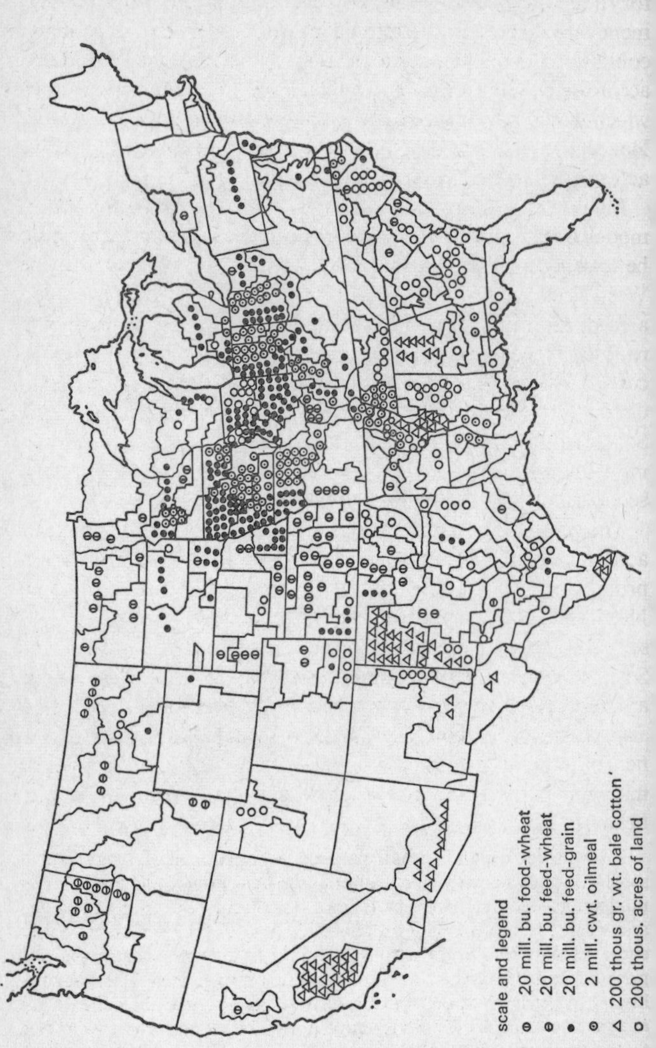

Figure 2 Optimum production of field crops, Model A. Demands – 1965 projection

for food wheat, feed grains, oilmeal, and cotton. Since requirements are defined as discrete quantities, rather than in terms of continuous demand functions, the results must be considered accordingly. The following results show the amounts of crops which would be produced, given the objective function as stated. Hence, the results also suggest the location and extent of surplus acreage devoted to the specified field crops.

Figure 2 shows the regional pattern of production arising from model A. As compared to the present, wheat production would be lessened in the marginal areas of the Great Plains and the West. Feed grain acreage could shrink into a more concentrated area of the Corn Belt, and into other advantageous but smaller regions. Soybean production would be held more or less to the current spatial distribution by models A and B (but not by C). Cotton acreage and production would shift greatly from the Southeast to the Southwest. Some wheat production would move into the Southeast[2] and some feed grains into the eastern seaboard of the Appalachian area.

The results presented in Figure 2 suggest a potential surplus acreage of land which might be withdrawn from grain and cotton production. This surplus acreage is concentrated in Montana, New Mexico, Texas, southeastern Colorado, Oklahoma, and selected regions of the Southeast, Appalachian, and Eastern Seaboard areas. Given the production coefficients as measured and the regional price structure assumed, the production pattern shown is the economically efficient one for producing the specified output mix. (It is not implied, however, that this output mix would be produced, in the short run, as a result of farmer behavior.)

2. The specification of wheat production in the Southeast may be due somewhat to inappropriateness of production coefficients and prices. First, production coefficients were based on a few observations from the limited area where wheat is presently grown in the region. They may not represent an entire region which might move into wheat production. Second, regional price differentials may be 'in the wrong direction' for this area. This area has a history for garlicky wheat, which is heavily discounted at the mills. In the analysis, however, United States prices for average grade wheat were used to construct price differentials, so that even though these prices were too high in terms of quality of the crop produced in the area, wheat would be included for the Southeast in the programming results, when actually it would be produced more efficiently elsewhere.

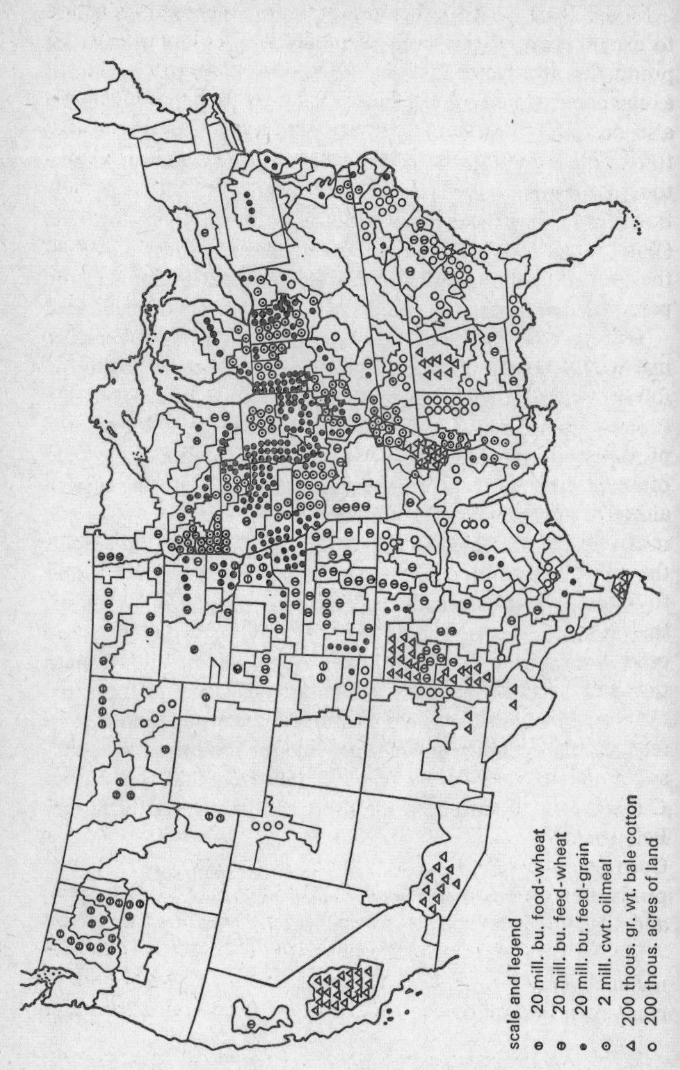

Figure 3 Optimum production location of field crops with oilseed and feed grain substitutions in consumption, Model B.

Model B permitted either feed grain or oilmeal production to expand beyond the levels of model A. Minimum-cost feed production is achieved by substituting feed grains for soybeans, as compared to model A (see Figure 3). Some other minor changes also occur in the regional pattern from model B as compared to A. Feed grain production is increased in region 31 (Indiana) to replace soybean production in this region. Feed grain production decreases in region 55 (South Dakota) and region 36 (Wisconsin). Wheat output increases in these two regions because the programmed price of feed grains declines slightly as compared to model A.[3]

The price of feed is lower under B than under A because the marginal units of feed are produced at lower costs. Model A forces a specific proportion of oilmeal into the production requirements, the resulting oilmeal component of total feed production thus being greater. Both models allow the production of soybeans only as a part of a feed grain rotation, so that although some regions may be able to produce soybeans at a relatively low cost, the cost of feed grain production included in the 'fixed' rotation may be relatively high. Therefore, if the total requirement of oilmeal is so high in a particular region that it must be selected to produce soybeans, the marginal quantities of feed grains (which are tied to the soybean production by rotation) will have relatively high unit cost.

Model C allows soybeans to be produced independently of feed grains. In other words, soybeans can be produced under a system of continuous cropping. As indicated for the results of C in Figure 4, soybean production is not specified for the Corn Belt by this model. The soybean production specified for the Corn Belt under the B solution is replaced, under C, by soybean production in region 19 (Mississippi) and region 64 (Nebraska) and additional production in region 21 and regions 60 and 65.

Since complete regional specialization in soybeans is allowed under model C, soybeans and feed grains for national requirements can be produced on fewer acres. Feed grains recede, in

3. The term 'programmed price' refers to the marginal supply price obtained by the dual programming solution. These prices determine whether a region will be selected to produce, and which activity within the region will be selected.

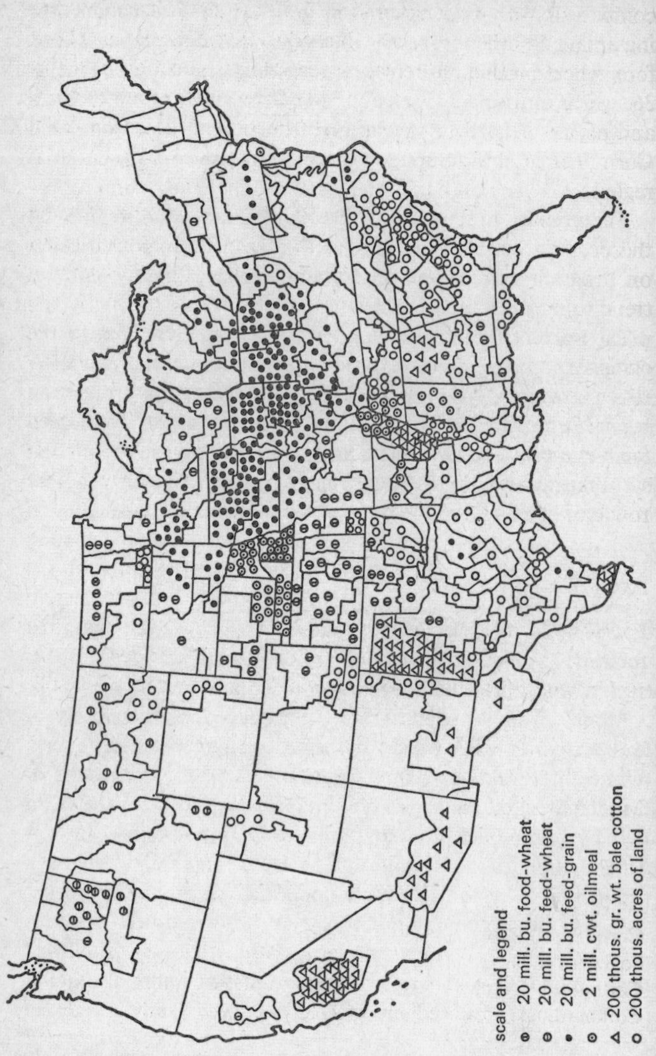

Figure 4 Optimum production location of field crops with regional soybean specialization and oilseed and feed grain substitution in consumption, Model C

comparison with models A and B, from the less efficient producing area, and other crops are allowed to take their place. Therefore, wheat production replaces feed grain production in region 36, which in turn substitutes for wheat production in region 53 and regions 12 and 13. Additional feed grain production in the Corn Belt in this solution replaces feed wheat production in region 93.

The primary purpose in application of model C was to show the possible impact of a greater degree of crop specialization on the optimum regional production and land use pattern. A trend towards greater specialization in field crops has been taking place over recent years. In the Corn Belt, for example, the relative acreages of corn and soybeans have been increasing, while the relative acreages of oats, forages, and other minor crops have been decreasing. This trend may become accelerated as more farmers learn that oats are a low return crop and as they find satisfactory ways to handle erosion problems under continuous row cropping (i.e., exclusion of meadow or hay rotations).

Average National Supply Prices

Each model solution gives rise to national average supply prices for each crop. These are presented in Table 2 to illustrate the structures arising from the regional production patterns depicted

Table 2
United States Average Supply Prices for Crops under the Three Models Where Land Costs Are Not Included in Coefficients

Crop	*A*	*B*	*C*
Wheat ($/bu.)	1·33	1·33	1·21
Corn ($/bu.)	1·09	1·09	0·95
Soybeans ($/bu.)	2·10	2·10	1·35
Cottonseed ($/ton)	39·15	39·15	28·26
Cotton ($/bale)	95·31	95·31	98·10

in Figures 2–4 and the particular characteristics of the data and the individual models. In general, these prices represent average marginal cost supply prices derived under the conditions and assumptions of the programming models and solutions. They cover all estimated costs except those for the land resource. Because regional differences are inherent in the models, imputed prices in individual regions are higher or lower than the averages given in the table. (If space had permitted, the implied regional prices also could have been presented.)

Basically, the individual commodity prices (Table 2) represent averages of regional prices as they might arise in a perfectly competitive situation and under the conditions assumed in the models. These average supply prices are lower than prevailing government price supports over recent years. However, they do not represent general equilibrium price quantities since, for the programming models, it was assumed that certain quantities (see Table 1) would be consumed if they were produced. (We did estimate disappearance with a general equilibrium price solution in mind: specifically with wheat at $1·39, corn at $1, soybeans at $1·80, and cotton at $120 per bale.)

The average supply prices (costs) derived for model A in which demands are discrete (see Table 2) are slightly higher, except for cotton, than those assumed for the consumption requirements used in the national restraints. Hence, both of two conditions would prevail: Actual demand would be slightly lower at the programmed supply prices than the demand requirements used in establishing national restraints. Production would be slightly higher than allowed by our restraints under an actual equilibrium solution.

Supply costs under model B, where feed grain can be substituted for oilmeal and expanded to meet total feed requirements, change so slightly from those of A that differences are not apparent in the rounded prices of Table 2. However, use of model C for soybean activities which can be grown outside of rotation allows a large change in cropping patterns and supply prices. All supply costs (prices) except cotton decline markedly for C as compared to B. These prices decline because soybeans and feed grains are produced singly in areas where they have the greatest absolute or comparative advantage. Or, in other words, feed grains and soybeans are produced separately in the regions where they have the lowest cost for the collective over-all model. With the added degree of specialization permitted in model C, the land base used for soybeans and feed grains shrinks. Consequently, land is released for other uses. This released land is more productive for wheat than are other areas used for wheat in the solution for model B. When this land is used for wheat production under C, the supply price of wheat also declines. The supply cost of cotton increases under model C as compared with B because of the decline in price of soybeans; the lowered soybean price forces a reduction in the price of cottonseed (and oilmeal from it), creating a cost-income deficit that must be made up by higher prices for cotton lint as a joint product with cotton seed.

Limitations of the Analysis

The results presented above are conditioned by resource and data deficiencies. Had research funds and time not been limited, more realistic models could have been constructed. These models would have included more crop activities and also livestock activities (studying feeder stock and slaughter animals, as well as market livestock).

Because we were not able to incorporate more detail and non-linear and non-discrete objective functions at the present, the results presented have some obvious limitations in exact specification of equilibrium quantities. However, as broad indicators of the competitive positions of particular areas in field crop production, the results are considered to be adequate and helpful in suggesting (a) adaptations in spatial patterns of crop production and (b) the general level of surplus cropland or producing capacity for particular crops. Additional time and refinements can allow estimates of even greater use in public policy and planning. The current study represents a necessary step in attaining these more refined estimates. The next step will include livestock activities and regional demand functions.

References

EGBERT, A. C., and HEADY, E. O. (1961), *Regional Production in Grain Production: A Linear Programming Analysis*, U.S. Department of Agriculture Tech. Bul. no. 1241, Washington.

HEADY, E. O., and EGBERT, A. C. (1959), 'Programming regional adjustments in grain production to eliminate surpluses,' *J. Farm Economics*, vol. 41, no. 4, pp. 718–33.

Part Three Applications of Other Programming Techniques

The resource allocation situations described in this part are all ones in which some of the assumptions needed for a successful linear programming solution break down. Various techniques are introduced depending on the extent to which the linear programming structure is changed.

The first two papers illustrate two techniques of non-linear programming: in the first, quadratic programming is applied to a problem of finance, while the second one solves an ore purchasing problem by means of separable programming. Probabilistic programming is introduced by Albach's paper, which describes a rather unusual long-range planning situation. This example is formulated as one of chance constrained programming. Certain of the structural variables are now random variables with known probability distributions and the constraints are permitted to be violated at specified probability levels. The objective function is also stochastic and it is the expected value of this function which is optimized. Zero-one programming is introduced and then combined with chance constrained programming in the following paper on the design of hospital menus.

The paper by Mellor is centred on a specific problem, job shop scheduling, rather than on any one technique. Several possible techniques are mentioned including integer programming, branch and bound, and the use of heuristics. A case study involving this latter approach is given in the next paper, and the final paper describes another heuristic programme applied to a further important problem, the location of warehouses.

6 N. R. Paine

A Case Study in Mathematical Programming of Portfolio Selections

N. R. Paine, 'A case study in mathematical programming of portfolio selections', *Applied Statistics*, vol. 1, 1966, pp. 24–36.

Introduction

In this section a synopsis of the E, V theory is presented.

Return is defined to be any measure of performance over time. Each investor may define return in the light of his particular status and objectives.

According to the E, V theory, the return on any portfolio constructed from a list of n securities is governed by a joint probability distribution with expected value E and variance V. The investment in a security on the list may be zero but not negative. An efficient portfolio constructed from the list of n securities – with expected return E_0 and variance of return V_0 – is a portfolio such that:

1. Any other portfolio constructed from the same security list, and with expected return E_0, has a variance of return greater than V_0, and
2. Any other portfolio constructed from the same security list, and with variance of return V_0, has an expected return less than E_0.

The E, V theory assumes that the investor will, after careful deliberation, desire to invest in some efficient portfolio rather than in some portfolio that is not efficient. Such an assumption appears reasonable enough for many investors.

Several algorithms for determining the set of all efficient portfolios have been developed (Markowitz, 1959). All algorithms that determine the set of efficient portfolios determine one by one all so-called corner portfolios. The corner portfolios are special efficient portfolios which describe the boundaries, so to speak, of the set of all efficient portfolios. The corner portfolio determined first we may call Corner Portfolio No. 1; the corner portfolio determined next, Corner Portfolio No. 2; etc. Corner Portfolio No. k is said to be adjacent to Corner Portfolio No. $(k - 1)$ and to Corner Portfolio No. $(k + 1)$. For any portfolio let X_i be the fraction of funds invested in the ith security. Then any portfolio is a vector $(X_1, X_2, X_3, \ldots, X_n)$. We require $X_i \geqslant 0$. Obviously $\sum_{i=1}^{n} X_i = 1$. The set of all efficient portfolios consists of all corner portfolios and all linear interpolation between all pairs of adjacent corner portfolios. Hence the corner portfolios may be said to describe the 'boundaries' of the set of all efficient portfolios.

There are two basic ways in which to arrive at probability distributions that purport to govern the performance of portfolios:

1. Probability distributions may be derived from probability beliefs (Markowitz, 1959). (This approach we may call the probability beliefs approach to the E, V theory.)
2. Probability distributions may be estimated by using past performance data for securities. (This approach we may call the historical approach to the E, V theory.)

A Useful Model

In this section a model used in connexion with the case study is described.

Suppose that the return on security i is

$$r_i = a_i + b_i I + u_i$$

where I is the value of some market index, a_i and b_i are two constants, and u_i is a random deviation, and suppose that the return on security j is similarly defined.

Let var stand for variance; cov, covariance; and expt, the expected value of. Suppose $\text{cov}(u_i, u_j) = 0$, $\text{cov}(I, u_i) = 0$, $\text{cov}(I, u_j) = 0$, $\text{expt}(u_i) = 0$, and $\text{expt}(u_j) = 0$. Then it may be shown that

$$\text{var}(u_i) = \sigma_{ii} - b_i^2 \, \text{var}(I),$$

where σ_{ii} is the variance of return for the ith security (Markowitz, 1959).

It may be shown that for any portfolio

$$V = \mathbf{X}'\mathbf{C}\mathbf{X}$$

where $\mathbf{X}' =$ the transpose of $\mathbf{X}$,

$$\mathbf{C} = \begin{bmatrix} \text{var}(u_1) & \cdots & 0 & 0 \\ \vdots & & \vdots & \vdots \\ 0 & \cdots & \text{var}(u_n) & 0 \\ 0 & \cdots & 0 & \text{var}(I) \end{bmatrix}$$

and $\mathbf{X}$ is a column vector with elements $X_1, X_2, \ldots, X_n, XI$,

where $$XI = \sum_{i=1}^{n} b_i X_i.$$

For any portfolio

$$E = \sum_{i=1}^{n} \mu_i X_i,$$

where $\mu_i =$ expected return of ith security.

Description of the Case Study

We shall assume that the model of the previous section provides an adequate portrayal of the performance of securities. Although the assumptions underlying the model may not always be entirely realistic, the model provides great computational advantages in that it eliminates the need for computing all possible covariances of return between pairs of securities. (From probability theory we know that $V = \sum_{i=1}^{n} \sum_{j=1}^{n} X_i X_j \sigma_{ij}$. The model may be utilized to derive from this formula the formula $V = \mathbf{X}'\mathbf{C}\mathbf{X}$, which does not involve all covariances.) In many instances the computational advantages of the model may far outweigh any theoretical objections.

The case study utilized the stocks in the portfolio of Founders Mutual Fund, an open-end mutual fund. Founders Mutual was created in 1938 under an agreement of trust between a trustee bank, a sponsor, and the investors who purchase beneficial interests in the trust. The sponsor is Founders Mutual Depositor Corporation. Founders Mutual follows the practice of investing equal dollar amounts in each of forty common stocks when funds are available for investment. The stocks represent issues of leading companies in important American industries. Founders Mutual is not a fixed trust, however, for the sponsor has the right to eliminate any stock at any time. The addition of a new stock requires the approval of a majority of the investors of the fund. As of the end of 1959, no stocks had been added to or dropped from the portfolio, and a policy of retaining stock received as the result of a stock split or stock dividend, but of selling stock rights and warrants had been followed.

For the period 1946–54 inclusive – a period covering approximately two American business cycles – the return on each stock was assumed to be governed by one probability distribution. For the period 1955–9 inclusive – a period covering approximately one American business cycle – the return on each stock was assumed to be governed by another probability distribution. Portfolios selected on the basis of parameters estimated from the earlier period will be evaluated in terms of their performance in the later period.

For each year of the period 1946–59 inclusive the return on each stock was computed. Roughly speaking, return for a year for a stock was defined to be the sum of dividends plus capital appreciation expressed as a percentage of market value at the first of the year. A precise definition is given later.

In defining return for a stock it was decided to make the following assumptions:

1. If a property dividend in the form of the stock of another company occurs, the stock received is sold at the average of the high price and the low price on the payment of the dividend. Ignore brokerage fees and commissions, transfer taxes, etc.

2. Any stock rights received or to be received are sold at the

average of the high price and the low price for the first day for which *The Commercial and Financial Chronicle* published quotations for the rights unless the rights are being sold on a 'when issued' basis in year t and the end of year t occurs when the stock is still selling with rights attached, in which case the rights are sold at the average of the high price and the low price for the first trading day of year $t + 1$. Ignore brokerage fees and commissions, transfer taxes, etc.

3. Stock received as the result of a stock split is retained until all holdings of the stock in question are liquidated.

4. With one exception (mentioned later), stock received as the result of a stock dividend is retained until all holdings of the stock in question are liquidated.

It was necessary to assume something with regard to the matters covered by these four assumptions. These assumptions seem to be as reasonable as any, do not involve any conflicts with the basic policy of Founders Mutual with regard to stock dividends and stock rights and warrants, and insure uniformity of treatment of various items arising for consideration.

The return on any stock for year t was defined as follows:

Assume that a certain number of shares of the stock is purchased at the closing price of year $t - 1$. (The number of shares is assumed to be any number such that no fractional shares are received during year t as the result of a stock split, a stock dividend, or a property dividend in the form of the stock of another company.) Assume that holdings of the stock are liquidated at the closing price of year t. Ignore brokerage fees and commissions, transfer taxes, etc. Then the return for the stock for year t is defined to be the decimal equivalent of a fraction which has as its denominator the total amount invested at the close of year $t - 1$ and which has as its numerator the following:

1. Proceeds from liquidation of holdings at close of year t, plus
2. Proceeds from cash dividends in year t, plus
3. Proceeds from sale of property dividends in the form of stock of another company in year t, plus
4. Proceeds from the sale of stock rights in year t, plus
5. Proceeds received in year $t + 1$ by virtue of ownership of the stock in year t, minus
6. Total amount invested at close of year $t - 1$.

A measure of performance such as this would seem to be excellent for a tax-free institution. Other investors might feel that such a measure is reasonably adequate or might want to define return in a somewhat different fashion.

If a stock dividend is received in year $t + 1$ by virtue of ownership in year t, it is assumed that the stock dividend is converted into cash at the average of the high price and the low price on the payment date of the stock dividend. Ignore brokerage fees and commissions, transfer taxes, etc. This assumption concerning the receipt of a stock dividend in year $t + 1$ constitutes the one exception to the previously mentioned assumption that stock received as the result of a stock dividend is retained until all holdings of the stock in question are liquidated.

For the period 1946–54 inclusive the following computations were made:

1. For each stock an estimate, $\hat{}_i$, of the expected value, μ_i, of the probability distribution assumed to govern the return of the stock was obtained by taking the mean of the observed returns of the stock.

2. For each stock an estimate of the variance $\sigma_{ii}{}^2$, of the probability distribution assumed to govern the return on the stock was obtained from

$$\sigma_{ij}^2 = \frac{1}{N-1} \sum_{t=1}^{N} (Y_{it} - \mu_i)^2$$

where

Y_{it} = return on the ith stock during the tth year of the period

N = number of years in the period,

and $\acute{\mu}_i$ is as previously defined.

3. For each year a value for an index I was computed from

$$I_t = \frac{S_t - S_{t-1}}{S_{t-1}}$$

where S_t is the value of Standard and Poor's Composite Daily Stock Price Index at the close of year t, S_{t-1} is the value of the Standard and Poor index at the close of year $t - 1$, and I_t is the value of the index I for the tth year of the period. The mean, $\bar{I}$, of the I_t values was computed.

4. For each stock an estimate of b_i was computed from

$$\hat{b}_i = \frac{\sum_{t=1}^{N} (I_t - \bar{I})\ (Y_{it} - \hat{\mu}_i)}{\sum_{t=1}^{N} (I_t - \bar{I})^2},$$

where the symbols are as previously defined.

5. An estimate of var(I) was obtained from

$$\hat{\text{var}}(I) = \sum_{t=1}^{N} \frac{(I_t - \bar{I})^2}{N - 1},$$

where the symbols are as previously defined.

6. For each stock an estimate of var(u_i) was obtained from

$$\hat{\text{var}}(u_i) = \hat{\sigma}_{ii}^2 - \hat{b}_i^2\ \hat{\text{var}}(I)$$

where the symbols are as previously defined.

The computations resulting from 1, 4, 5 and 6 above are shown in Table 1. These figures were used in determining corner portfolios for the period 1946–54 inclusive. The computations involved in determining the corner portfolios were performed by an I.B.M. 7090 using the quadratic programming code developed by the RAND Corporation of Santa Monica, California (Wolfe, 1959). The corner portfolios are described in Table 2. Stocks not shown are not included in any efficient portfolio.

Note that no efficient portfolio contains more than nine securities. A very important practical implication here is that actual fund operators may overdo diversification. Of course, a very large fund may have to diversify simply because it is large; it may be unable to buy as much of one stock as it would like or to unload when it would like.

By reference to the E and $\sqrt{V}$ values for each corner portfolio an approximation to the so-called E, $\sqrt{V}$ curve – which shows $\sqrt{V}$ as a function of E for all efficient portfolios – may be drawn. See Figure 1. Point 1 refers to Corner Portfolio No. 1, etc. An E, $\sqrt{V}$ curve is made up of a series of segments of hyperbolas (Markowitz, 1959).

Any portfolio with $X_i = 0{\cdot}025$, $i = 1, \ldots, 40$, has $E = 0{\cdot}161$

Table 1
Key Values for the Period 1946–1964

i Security no.	Description	Estimates of μ_i	b_i	var (u_i)
1	Union Pacific R.R. Co.	0·167	0·8093	0·018 237
2	American Telephone and Telegraph Co.	0·057	0·3604	0·001 473
3	Insurance Company of North America	0·184	0·4008	0·024 903
4	Chrysler Corporation	0·128	0·9218	0·066 631
5	General Motors Corporation	0·228	1·6221	0·011 417
6	United Aircraft	0·256	1·7209	0·048 553
7	United States Gypsum Co.	0·186	1·5154	0·036 956
8	Air Reduction Co. Inc.	0·022	1·4247	0·014 782
9	Du Pont (E.I.) de Nemours and Co.	0·219	1·0186	0·015 133
10	Dow Chemical Co.	0·216	0·6489	0·035 125
11	Monsanto Chemical Co.	0·186	0·2683	0·087 993
12	Union Carbide Corp.	0·154	0·3291	0·005 467
13	Radio Corporation of America	0·200	1·7417	0·058 943
14	Continental Can Co. Inc.	0·129	0·8842	0·026 753
15	Owens-Illinois Glass Co.	0·104	0·7739	0·025 870
16	General Electric Company	0·202	1·1562	0·013 741
17	Westinghouse Electric Corp.	0·171	1·2173	0·025 237
18	Corn Products Refining Co.	0·088	0·2043	0·013 345
19	Procter and Gamble Company	0·154	0·7738	0·016 203
20	Caterpillar Tractor Co.	0·208	1·6719	0·037 370
21	Ingersoll-Rand Company	0·188	1·2317	0·009 535
22	International Harvester Co.	0·099	1·0325	0·015 211
23	Aluminium Co. of America	0·186	0·7819	0·050 333
24	American Smelting and Refining Co.	0·174	1·4679	0·023 779
25	International Nickel Company of Canada Ltd	0·138	1·5349	0·018 303
26	Kennecott Copper Corp.	0·198	1·3073	0·032 357
27	Continental Oil Co.	0·237	0·9612	0·028 269
28	Standard Oil Co.	0·236	1·2096	0·039 708
29	Texaco, Inc.	0·207	1·1252	0·010 395
30	Sears Roebuck and Co.	0·143	0·4092	0·022 540
31	F. W. Woolworth Co.	0·054	0·4144	0·005 444
32	United States Rubber Co.	0·194	1·4481	0·025 018

i Security no.	Description	Estimates of μ_i	b_i	*var* (u_i)
33	Inland Steel Co.	0·195	1·4605	0·045 936
34	United States Steel Co.	0·233	1·8041	0·029 269
35	Liggett and Myers Tobacco Company 'B'	0·023	0·0108	0·008 665
36	R. J. Reynolds Tobacco Company 'B'	0·076	0·1145	0·020 552
37	Pullman, Inc.	0·111	1·4180	0·032 100
38	Eastman Kodak Co.	0·142	0·8024	0·010 406
39	International Business Machines Corp.	0·253	1·0105	0·043 307
40	Parke, Davis and Co.	0·086	0·6956	0·085 097

Estimate of var (I): 0·029 708

and $\sqrt{V} = 0{\cdot}173$ and is represented in Figure 1 by point F. Any such portfolio is a portfolio constructed in accordance with a policy somewhat like that followed by Founders Mutual Fund.

From Figure 1 we see that the efficient portfolio directly below point F is Corner Portfolio No. 14 and the efficient portfolio with a $\sqrt{V}$ value equal to the $\sqrt{V}$ value of point F has an E value of approximately 0·232. This portfolio, falling just below Corner Portfolio No. 7 on the E, $\sqrt{V}$ curve, will be referred to as Portfolio No. 7′. Portfolio No. 7′ is composed of the following securities (Markowitz, 1959).

Security No.	*Fraction of funds invested in the security*
3	0·04229204
10	0·22490017
11	0·05949206
27	0·30426821
28	0·04761534
39	0·32143218

The same values shown in Table 1 for 1946–54 were computed for 1955–9. See Table 3. The values in Table 3 were used in computing E, V and $\sqrt{V}$ for 1955–9 for each corner portfolio of

Table 2

Corner Portfolios for a Forty Security Analysis

Security no.	*Fraction of funds invested in each security*										
	Corner portfolio no.										
	1	*2*	*3*	*4*	*5*	*6*	*7*	*8*	*9*	*10*	*11*
2											
3								0·08458408	0·22052951	0·21032598	0·19872077
6	1	0·26467222	0·11015021	0·08029011							
9									0·03659149	0·02176290	
10					0·17293138	0·18618888	0·22485508	0·22494527	0·21934936	0·19123044	0·15948344
11							0·04993070	0·06905343	0·09970969	0·09226120	0·08419893
12										0·15000030	0·31213124
18											
27			0·31841046	0·35043752	0·35017367	0·34807338	0·32340407	0·28513234	0·21160241	0·17233416	0·12688569
28				0·02957021	0·05906927	0·05912545	0·05404161	0·04118907	0·01188517		
30											
31											
35											
36											
39		0·73532778	0·57143933	0·53970216	0·41782569	0·40661228	0·34776854	0·29509581	0·20033236	0·16208502	0·11857992
E	0·256	0·254	0·248	0·247	0·240	0·239	0·235	0·229	0·218	0·207	0·195
V	0·136533	0·069492	0·051804	0·049571	0·038604	0·038068	0·033676	0·028511	0·021991	0·017188	0·013092
$\sqrt{V}$	0·370	0·264	0·228	0·223	0·196	0·195	0·184	0·169	0·148	0·131	0·114

Table 2 (continued)

Security no.	Fraction of funds invested in each security									
	Corner portfolio no.									
	12	*13*	*14*	*15*	*16*	*17*	*18*	*19*	*20*	*21*
2									0·24100322	0·29331092
3	0·18785490	0·13583582	0·13168707	0·12825672	0·12336084	0·10471303	0·05516168	0·04570689	0·01524094	0·00710702
6										
9										
10	0·13179395	0·07057464	0·06635152	0·06261515	0·05780241	0·03798317				
11	0·07756970	0·05743742	0·05581286	0·05456049	0·05275798	0·04639803	0·02739324	0·02388760	0·01407545	0·01151758
12	0·43647761	0·43016267	0·42630627	0·42289355	0·41538255	0·38624128	0·26361771	0·23978875	0·14168758	0·11517073
18					0·01081605	0·05051808	0·10040190	0·10916657	0·10725423	0·10622142
27	0·08578971	0·00497696								
28										
30		0·28804817	0·30203964	0·31053365	0·31558913	0·31976135	0·25946045	0·24349503	0·09186679	0·04880379
31										0·01673977
35							0·20490184	0·24258381	0·29236136	0·30457270
36			0·00903625	0·01637222	0·02429104	0·05438506	0·08906317	0·09537135	0·09651042	0·09655607
39	0·08051412	0·01296431	0·00876639	0·00476822						
E	0·185	0·163	0·161	0·159	0·157	0·150	0·113	0·107	0·078	0·069
V	0·010655	0·006743	0·006485	0·006319	0·006061	0·005397	0·003290	0·003122	0·002726	0·002707
$\sqrt{V}$	0·103	0·082	0·081	0·079	0·078	0·073	0·057	0·056	0·052	0·051

Table 3
Key Values for the Period 1955–59

i Security no.	*Description*	*Estimates of* μ_i	b_i	$var(u_i)$
1	Union Pacific R.R. Co.	0·079	1·2883	0·012 381
2	American Telephone and Telegraph Co.	0·138	0·6275	0·008 537
3	Insurance Company of North America	0·120	0·8174	0·023 547
4	Chrysler Corporation	0·055	0·5403	0·047 954
5	General Motors Corporation	0·195	1·5206	0·003 240
6	United Aircraft	0·075	0·9738	0·071 366
7	United States Gypsum Co.	0·217	0·7115	0·038 145
8	Air Reduction Co. Inc.	0·268	0·9983	0·021 956
9	Du Pont (E.I.) de Nemours and Co.	0·149	0·8326	0·025 799
10	Dow Chemical Co.	0·222	1·1247	0·009 835
11	Monsanto Chemical Co.	0·169	0·5755	0·060 519
12	Union Carbide Corp.	0·161	0·9696	0·002 482
13	Radio Corporation of America	0·213	1·3704	0·056 102
14	Continental Can Co. Inc.	0·096	1·0129	0·022 366
15	Owens-Illinois Glass Co.	0·204	1·0341	0·007 243
16	General Electric Company	0·200	0·5056	0·004 509
17	Westinghouse Electric Corp.	0·122	0·2454	0·074 157
18	Corn Products Refining Co.	0·215	0·6384	0·044 323
19	Procter and Gamble Company	0·171	0·2541	0·012 886
20	Caterpillar Tractor Co.	0·286	1·4664	0·051 379
21	Ingersoll-Rand Company	0·167	1·1489	0·026 903
22	International Harvester Co.	0·149	1·2966	0·037 816
23	Aluminium Co. of America	0·285	2·1614	0·054 581
24	American Smelting and Refining Co.	0·121	1·1284	0·020 273
25	International Nickel Company of Canada Ltd	0·207	1·0602	0·039 882
26	Kennecott Copper Corp.	0·072	1·1034	0·014 077
27	Continental Oil Co.	0·158	1·4167	0·029 850
28	Standard Oil Co.	0·123	0·7523	0·028 201
29	Texaco, Inc.	0·210	0·8376	0·014 736
30	Sears Roebuck and Co.	0·227	1·4894	0·022 777
31	F. W. Woolworth Co.	0·133	0·9315	0·040 858

i Security no.	Description	Estimates of μ_i	b_i	$var(u_i)$
32	United States Rubber Co.	0·169	1·3951	0·021 691
33	Inland Steel Co.	0·244	1·9565	0·063 542
34	United States Steel Co.	0·335	2·1869	0·018 821
35	Liggett and Myers Tobacco Company 'B'	0·146	0·3841	0·005 109
36	R. J. Reynolds Tobacco Company 'B'	0·294	0·4537	0·011 979
37	Pullman, Inc.	0·106	1·2378	0·012 350
38	Eastman Kodak Co.	0·303	0·4721	0·022 203
39	International Business Machines Corp.	0·409	0·6505	0·079 379
40	Parke, Davis and Co.	0·373	0·9105	0·085 979

Estimate of var (I): 0·041 978

1946–54, for Portfolio No. 7′, and for a portfolio with $X_i = 0{\cdot}025$, $i = 1, \ldots, 40$. The results are tabulated in Table 4 and are shown graphically in Figure 2. In Figure 2 point 1 refers to Corner Portfolio No. 1, etc.

Results and Conclusions

We may now consider two ways in which the E, V theory may be put to practical use by portfolio managers.

A. The following hypothetical example illustrates one way in which a portfolio manager might utilize the E, V theory.

Suppose the portfolio manager maintained for 1946–54 inclusive a portfolio composed of the forty stocks in Founders Mutual Fund. He might then compare the performance of his portfolio with the performance of some of the efficient portfolios for 1946–54. (All of the efficient portfolios may be determined from Table 2.) The comparisons thus made would give him some basis for evaluating and improving his work.

B. Let us now consider an example which illustrates another way in which the E, V theory might be utilized.

Suppose an investor has a portfolio in 1946–54 corresponding to point F in Figure 1 (i.e. the investor follows an investment policy similar to that of Founders Mutual Fund for 1946–54).

Suppose that for 1955–9 inclusive the investor chooses a portfolio that was efficient in 1946–54. Is the change in investment policy beneficial? It will be argued that unless the investor chose an efficient portfolio from two rather small portions of the E, $\sqrt{V}$ curve of Figure 1, the investment policy change would be beneficial.

Table 4
Performance of Key Portfolios for the Period 1955–9

Portfolio		E	V	$\sqrt{V}$
Corner Portfolio No.	1	0·075	0·111 173	0·333
	2	0·321	0·070 664	0·266
	3	0·292	0·066 126	0·257
	4	0·286	0·064 996	0·255
	5	0·272	0·060 463	0·246
	6	0·270	0·060 129	0·245
	7	0·258	0·055 993	0·237
Portfolio No. 7′		0·250	0·053 569	0·231
Corner Portfolio No.	8	0·243	0·051 382	0·227
	9	0·214	0·044 973	0·212
	10	0·204	0·043 100	0·208
	11	0·192	0·041 317	0·203
	12	0·182	0·039 945	0·200
	13	0·182	0·052 469	0·229
	14	0·183	0·052 812	0·230
	15	0·184	0·053 101	0·230
	16	0·185	0·052 862	0·230
	17	0·190	0·050 657	0·225
	18	0·190	0·035 540	0·189
	19	0·190	0·033 016	0·182
	20	0·175	0·020 688	0·144
	21	0·171	0·018 328	0·135
A Portfolio with $X_i = 0{\cdot}025$		0·190	0·042 739	0·207

Suppose that

$$U = f(E, \sqrt{V}),$$

$$\frac{\partial U}{\partial E} > 0$$

and

$$\frac{\partial U}{\partial \sqrt{V}} < 0,$$

where U represents utility to the investor. That is, we make the fairly reasonable assumptions that utility to the investor depends

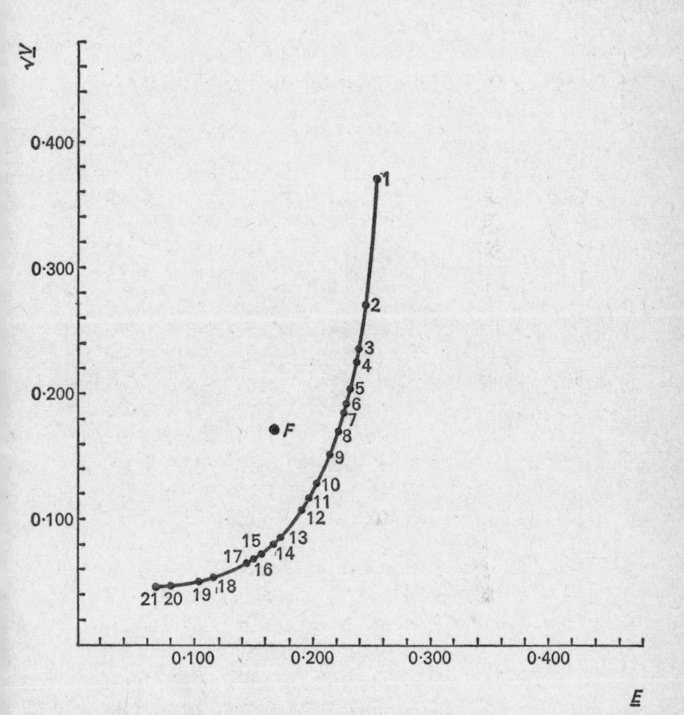

Figure 1 An E, $\sqrt{V}$ curve for a forty security analysis

on E and $\sqrt{V}$ and that the investor likes E and dislikes $\sqrt{V}$. Then it is clear that:

1. Any point in Figure 2 with $E > 0{\cdot}190$ and $\sqrt{V} < 0{\cdot}207$ (these values being the coordinates of point F) corresponds to a portfolio which is better than the portfolio associated with point F, and

2. Any point with $E < 0{\cdot}190$ and $\sqrt{V} > 0{\cdot}207$ corresponds

to a portfolio which is worse than that associated with point F. It can be seen from Figure 2 that the portfolio corresponding to point F is better than the portfolios corresponding to points 1, 13, 14, 15, 16 and 17 and is worse than the portfolios corres-

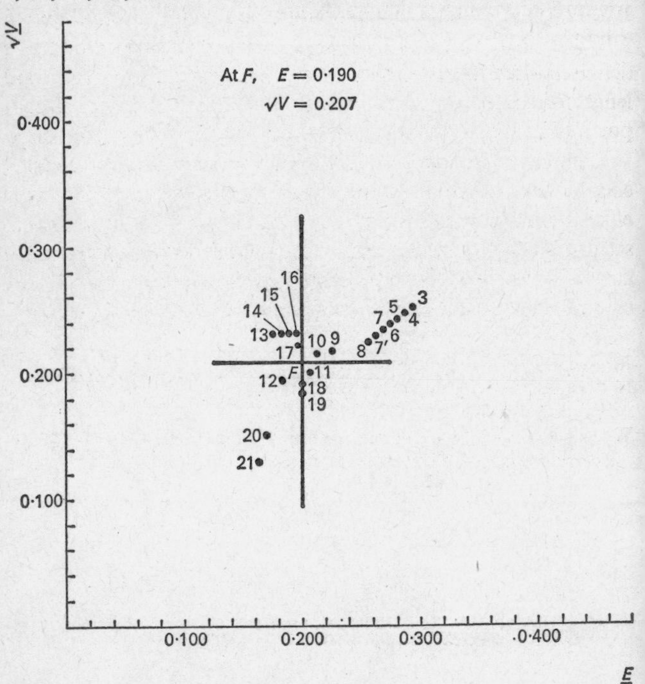

Figure 2 $E, \sqrt{V}$ values for the period 1955–9

ponding to points 11, 18 and 19. Much can be said for the argument that portfolios corresponding to all of the other points in Figure 2 except point 12 (i.e. points 2–10, 20–21) would be better for most investors than the portfolio corresponding to point F. For each of these points either E is greater (and $\sqrt{V}$ greater to a much lesser extent), or E is less (and $\sqrt{V}$ much less), than at point F. That is, if the investor moves from F to any of these points, he either:

1. Obtains a higher E at the expense of only a relatively slight increase in $\sqrt{V}$, or

2. Obtains a smaller $\sqrt{V}$ at the expense of only a relatively slight decrease in E.

At any rate, the portfolios corresponding to points 1 and 12–17 in Figure 2 represent portfolios definitely inferior in 1955–9 to the portfolio corresponding to point F. From Figure 1 it is clear that the portion of the E, $\sqrt{V}$ curve corresponding to definitely inferior portfolios is small in relation to the total length of the E, $\sqrt{V}$ curve. Thus the definitely inferior efficient portfolios are only a small proportion of all the efficient portfolios.

The above example certainly does not prove that an investor should always invest in the future in portfolios which were efficient in the past. However, if an investor is interested in securities which, he feels, will perform in the future approximately as they have in the past, then he may do well to consider very carefully investment in portfolios which were efficient in the past.

References

Markowitz, H. M. (1959), *Portfolio Selection: Efficient Diversification of Investments*, Wiley and Chapman and Hall.

Wolfe, P. (1959), 'The Simplex method for quadratic programming', *Econometrica*, vol. 27, p. 382.

7 E. M. L. Beale, P. J. Coen and A. D. J. Flowerdew

Separable Programming Applied to an Ore Purchasing Problem

E. M. L. Beale, P. J. Coen, and A. D. J. Flowerdew, 'Separable programming applied to an ore purchasing problem', *Applied Statistics*, vol. 14, 1965, pp. 89–101.

Introduction

This paper describes the use of Separable Programming to determine the optimum mix of raw materials (chiefly iron ore) to be converted to liquid iron in four blast furnaces. The iron ore is mostly imported, and is ordered annually through the British Iron and Steel Corporation (Ore) Ltd. Some, but not necessarily all, of the iron ore may be pre-processed in a sinter plant to form sinter, a material with better physical properties for iron-making in blast furnaces than the un-processed ore. The sinter plant is situated at Spencer Works, and its output may be used both in the two blast furnaces there and in the blast furnaces at Ebbw Vale Works. In the furnaces, liquid iron and slag are produced from iron ore, sinter and smaller quantities of other iron-bearing materials, together with coke and fluxes such as limestone and dolomite.

Formulation

Linear programming applied to blast furnaces

Although iron has normally to undergo further processing to become steel the iron analysis must be kept within fairly close limits for the sake of efficiency in the steel-making shop. The distribution of elements such as phosphorus and sulphur between the liquid iron and the liquid slag depends on such factors as contact time between iron and slag, temperature, amount of each element present and slag composition. The last two factors must be controlled through the mix of raw materials (known as the burden). This control is achieved through restrictions of the type:

Sulphur in burden must not exceed x per cent of total slag weight.

To ensure efficient furnace operation, the physical state of the raw materials charged must also be controlled. For example, production rate depends on, among other factors, the particle size distribution of the burden. Some of the ores contain a high proportion of small particles and are referred to as 'fine ores'. A typical restriction on the physical characteristics of the burden might be:

Fine ores must not exceed y per cent of total burden.

The available supply of many of the ores is restricted, either because of the mine production, or because only a limited amount is to be shipped to Britain, or made available to a particular company. There will therefore be market restrictions of the type:

Maximum amount of ore A is Z tons/year.

From this brief description it should be clear that the problem of choosing an optimum burden for a blast furnace is of the mathematical programming type. In practice, it can normally be set up as a linear programming problem.

The linear programming problem consists of an objective function to be maximized or minimized, a set of decision variables, and a set of restrictions (equalities or inequalities) on the values of the decision variables. In the blast furnace burdening problem, the objective function (to be minimized) is the total cost of the burden expressed as a function of the quantities of raw materials selected; the decision variables are these quantities of raw materials; and the restrictions are the physical, chemical and market limitations on the burden, as described above, together with equations specifying fuel requirements, required production rate, and other furnace limitations.

The simplex method

This is not an appropriate place to describe in detail the simplex method for solving linear programming problems, or the various modifications to it. Full details have been widely published; see, for instance, Vajda (1960). However, a brief formal description of the problem may be useful as a reminder, and to set the scene for the introduction of Separable Programming in a later section.

Formally, the problem is to find a set of non-negative variables x_i that minimize some linear *objective function* $C = \sum_{j=1}^{n} C_j x_j$, while satisfying the equations

$$\sum_{j=1}^{n} a_{ij} x_j = b_i \quad \text{for } i = 1, \ldots, m. \qquad \textbf{1.1}$$

Note that an inequality constraint can be turned into an equation and therefore put in the form (1.1) by adding a *slack variable* that must be non-negative representing the difference between the left and right hand sides of the inequality.

In outline the simplex method solves this problem as follows. The constraints (1.1) are solved for m variables in terms of the remaining ones, so that they read

$$x_k = a'_{k0} + \sum a'_{kj} x_j \quad \text{for } k = j_1 \ldots j_m\,, \qquad \textbf{1.2}$$

where the summation extends over all other variables. The variables on the left hand sides of those equations are called *basic variables*, and the set of such variables a *basis*. The variables on the right hand sides are called *nonbasic variables*. We now express the objective function C in terms of the nonbasic variables.

Corresponding to this way of expressing the problem we associate a trial solution obtained by putting all the nonbasic variables equal to zero. The values of the objective function and the basic variables are then given by the constant terms in the equations.

The nonbasic variables can be thought of as independent variables, and the basic variables as dependent variables. But we are going to solve the problem by an iterative procedure, in which the set of nonbasic variables will be continuously changing. It is therefore reasonable to give them this special name, nonbasic, instead of simply calling them independent variables.

Let us assume that our trial solution is *feasible*, i.e. that all the basic variables are non-negative. (The simplex method can be used to find a feasible solution by working with an auxiliary objective function representing the sum of the infeasibilities, but we cannot discuss all the details here.) We now want to find out whether it is the best possible solution. To find out, we look at the signs of the coefficients of the nonbasic variables in the expression for the objective function. If these are all non-negative then the problem is solved. Otherwise we take a variable with a negative coefficient, and increase its value keeping the other nonbasic variables constant and equal to zero. This will change the values of the basic variables, and our progress will be halted as soon as any basic variable reaches the value zero – since we cannot allow it to go negative. But at this stage, we can make the variable we are increasing basic (i.e. *introduce it into the basis*), in place of the one whose value is zero. We substitute for the new basic variable in terms of the nonbasic variables throughout the constraints and the objective function, and we are back in a similar position to the one we started from but with a smaller value of the objective function. So we can carry on making profitable changes of basis in this way until the problem is solved.

The simplex method can be described geometrically. The feasible region of the linear programming problem **1.1** is a polyhedron in $(n - m)$ dimensional space. The method starts from some vertex of the polyhedron and at each step looks for a neighbouring vertex with a better value of the objective function. When so such vertex exists an optimum solution has been found.

Formulation of the ore purchasing problem

Before the approach described in this paper was developed, the method of calculating burdens used was as follows. For Ebbw Vale, the analysis and quantity of sinter available was taken as given, and this sinter was treated as a possible input to the Ebbw Vale blast furnaces. For Spencer Works the maximum sinter available was taken to be the difference between sinter plant capacity and the sinter required by Ebbw Vale; separate costs and properties for sintered and unsintered ores were estimated and the computer programme determined the production level of the sinter plant and the mix of ores for sintering, and hence the sinter analysis, as well as the actual burdens for the Spencer Works blast furnaces.

The drawbacks of this method are that neither calculation can be performed first, since the inputs to each depend on the output from the other, and that the mix of ores for sintering ought to depend on the desiderata for both works, not just for Spencer Works. An iterative procedure might have been followed, starting from guessed values for, say, the sinter analysis and quantity available to Ebbw Vale, but there seemed no guarantee that such a procedure would converge, still less to an optimum. It was therefore decided to try to set up a model for both works combined as a mathematical programming problem, and see whether a technique was available to solve it.

We start with four classes of variables:

Inputs to sinter plant	x_{0k}
Inputs to Ebbw Vale blast furnace 1	x_{1k}
Inputs to Ebbw Vale blast furnace 2	x_{2k}
Inputs to Spencer Works blast furnaces	x_{3k}

(The two Ebbw Vale furnaces are treated separately because they produce iron to different specifications; the Spencer Works furnaces are, for burdening purposes, identical and may be treated as one.)

The objective function is, as usual, the sum of the costs of these raw materials, and though the normal difficulties arise in estimating the appropriate costs to use, the function remains linear. The complications arise with the constraints. We shall illustrate this by deriving the appropriate constraint for manganese in the Spencer Works blast furnaces. Here the constraint is of the form:

Manganese content of Spencer Works hot metal must be at least m per cent.

We must first determine the total manganese input to the furnaces; this is given by

$$M = \sum m_k x_{3k}, \qquad \textbf{1.3}$$

where m_k is the manganese content of the kth input. For a given production level H and a given distribution of manganese between hot metal and slag (Q per cent to hot metal) we then have

$$Q\frac{M}{H} \leqslant m. \qquad \textbf{1.4}$$

Inequality **1.4** is clearly linear since Q and H are constants.

What about **1.3**? For the unsintered ores, m_k is a constant, namely the manganese content of the ore expressed as a proportion. However, sinter is also an input variable, and both the quantity and its manganese content vary depending on which ores are chosen to make sinter and how much sinter is used at Spencer Works. So that the constraint **1.3** contains a term which is the product of two variables.

We first attempted to resolve this difficulty by treating the total manganese input from sinter as a variable x_{m3}. If the total manganese content of the sinter is x_m, then we have

$$x_m = x_{m1} + x_{m2} + x_{m3}, \qquad \textbf{1.5}$$

where x_{m1} and x_{m2} are the manganese inputs at the two Ebbw Vale furnaces.

x_m can now be defined by an equation of the form

$$x_m = \sum m_k x_{0k}, \qquad \textbf{1.6}$$

where as before m_k are the manganese contents of the ores and x_{0k} are the sinter input variables. However, if no further constraints than **1.5** and **1.6** were included, the programme would be able to choose the proportions of any two elements, say manganese and phosphorus, in the sinter going to Ebbw Vale, separately and these might be different from the proportions in the sinter for Spencer Works. In the extreme case, all the phosphorus might go to Ebbw Vale, with all the manganese to Spencer Works. So that additional restrictions are necessary to make sure that the ratio $x_{m1} : x_{m2} : x_{m3}$ is equal to the ratio of the sinter going to each furnace, $S_{01} : S_{02} : S_{03}$ say. When this is done the product terms reappear.

To define the problem more precisely, we introduce some more variables as follows:

S_0 is the total sinter make

S_{0j} is the sinter used at Furnace j ($j = 1$ for Ebbw Vale 1, $j = 2$ for Ebbw Vale 2, $j = 3$ for Spencer Works)

S_i is the total amount of chemical component i in the sinter make ($i = 1 \ldots 8$)

S_{ij} is the amount of chemical component i in the sinter sent to the indicated furnace

Then[1] the objective function to be minimized is of the form

$$\sum_k C_k (x_{0k} + x_{1k} + x_{2k} + x_{3k}) + C_0 S_0,$$

where the C_k are the costs of the raw materials, and C_0 the cost of sintering. The first group of constraints define the amount of hot metal H_j to be made at each furnace, i.e.

$$\sum_k a_{1k} x_{jk} + S_{1j} = H_j \quad \text{for } j = 1, \ldots, 3, \qquad \textbf{1.7}$$

where a_{1k} is the iron content of raw material k.

The next group of constraints limits the amounts of other chemical components in the burdens, i.e.

$$L_{ij} \leqslant \sum_k a_{ik} x_{jk} + S_{ij} \leqslant U_{ij} \quad \text{for } i = 2, \ldots, 8, j = 1, \ldots, 3. \quad \textbf{1.8}$$

Here L_{ij} and U_{ij} are the lower and upper bounds on component i and a_{ik} is the quantity of component i in raw material k.

There will be other linear constraints on the input materials as indicated above. We omit these to simplify the exposition. But we should indicate the constraints on the sinter which are as follows:

The total sinter make, and its chemical composition, are linearly related to the inputs to the sinter plant, i.e.

$$S_i = \sum_k r_{ik} x_{0k} \quad \text{for } i = 0, 1, \ldots, 8, \qquad \textbf{1.9}$$

where r_{ik} gives the yield on sintering of raw material k for component i, and over all ($i = 0$).

Also the materials in the sinter make must equal the materials in the sinter used at the different furnaces, i.e.

$$S_i = S_{i1} + S_{i2} + S_{i3} \quad \text{for } i = 0, 1, \ldots, 8. \qquad \textbf{1.10}$$

Finally, we have the non-linear constraints which ensure that the sinter sent to the different furnaces is homogeneous, i.e.

$$\frac{S_{i1}}{S_{01}} = \frac{S_{i2}}{S_{02}} = \frac{S_{i3}}{S_{03}} \quad \text{for } i = 1, \ldots, 8. \qquad \textbf{1.11}$$

1. The formulation given here is slightly simplified compared with the model used in practice but contains all the essential features. For instance, the costs of raw materials at the two works differed because of transport costs.

We have now formulated a typical non-linear programming problem, i.e. a linear programming problem with a certain number of non-linear constraints.

Computation

Separable programming

A powerful general method of solving such problems has been developed by Miller (1963). This method is not as well known as it deserves to be; perhaps because it is so simple mathematically that it may appear trivial to the theorist. We, therefore, describe the method in general terms before returning to our specific problem.

The method is known as 'Separable Programming', since it assumes that all the non-linear constraints can be separated out into sums and differences of non-linear functions of single variables. At first this assumption seems to severely restrict the usefulness of the method. But for reasons discussed later this is not really so.

Now suppose that we have some variable z, and we want to deal with a function $f(z)$. Suppose that the graph of $f(z)$ looks something like Figure 1.

We now replace this function by a piecewise linear approxi-

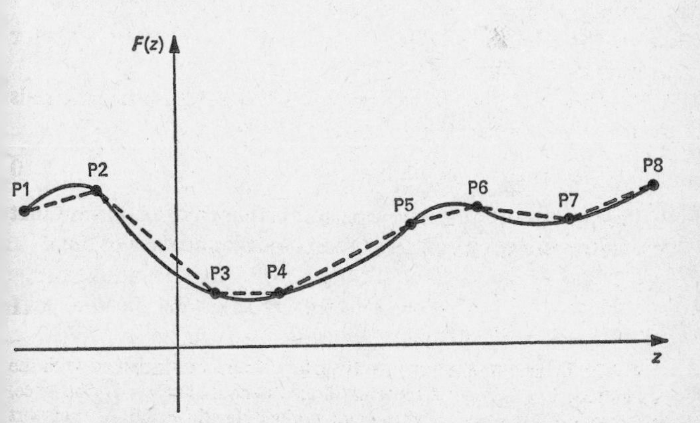

Figure 1 A piecewise linear approximation to a continuous function

mation based on a finite number of points. In the diagram we have taken 8 points $P_1 \ldots . P_8$.

Now let the co-ordinates of the 8 points be (a_i, b_i), and introduce 8 new non-negative variables $\lambda_1 \ldots \lambda_8$ and the equations

$$\lambda_1 + \ldots + \lambda_8 = 1, \qquad 2.1$$

$$a_1 \lambda_1 + \ldots + a_8 \lambda_8 = z, \qquad 2.2$$

$$b_1 \lambda_1 + \ldots + b_8 \lambda_8 = f(z). \qquad 2.3$$

We call these variables a single group of 'special variables' for reasons that will appear very shortly.

Let us consider some typical solutions of these equations.

If we put $\lambda_1 = 1, \lambda_2 = \ldots = \lambda_8 = 0$, then $z = a_1$ and $f(z) = b_1$ and we have the point P_1.

If we put $\lambda_1 = \frac{1}{2}, \lambda_2 = \frac{1}{2}, \lambda_3 = \ldots = 0$, then $z = \frac{1}{2}(a_1 + a_2)$ and $f(z) = \frac{1}{2}(b_1 + b_2)$ and we have a point half way between P_1 and P_2.

More generally, if we allow any two neighbouring special variables to take non-zero values, holding the other special variables of the group at zero, we will map out the piecewise linear approximation to $f(z)$ that we have agreed to use. On the other hand if we put say $\lambda_1 = \frac{1}{2}$ and $\lambda_3 = \frac{1}{2}, \lambda_2 = \lambda_4 = \ldots = 0$ we have a point midway between P_1 and P_3 which is not a valid one.

Now in some problems we know beforehand, perhaps from convexity considerations, that such inadmissible combinations of special variables cannot occur in an optimal solution even if we take no special steps to exclude them. In these circumstances we do not need to use Separable Programming.

Separable Programming is a method of reaching a local optimum, which may possibly not be a global optimum to a non-convex problem by taking special steps to exclude inadmissible combinations of special variables. This is very easy if we are using the simplex method. At each iteration we have to see if we can improve the value of the objective by increasing one of the nonbasic variables (whose value is zero at the current trial solution), keeping the other nonbasic variables equal to zero. We then introduce this variable into the basis, as indicated in part one. In separable programming all we have to do is to restrict the set of variables to be considered as candidates for entering the basis. If two special variables of a group are already in the basis we do not consider allowing a third. And if one is already in the basis we consider only its neighbours.

And that is all there is to the theory of separable programming. A number of technical points have to be considered when one comes to apply it, and we illustrate some of these with our particular application.

Application of separable programming – product terms

A feature of our problem that applies to many non-linear programming problems is the presence of product terms. Now the expression $u_1 u_2$ is not a non-linear function of a single variable, so it might appear that it was not amenable to separable programming. If that were true, then separable programming would be of very limited value; but fortunately it is not true.

We note that

$$u_1 u_2 = \left(\frac{u_1 + u_2}{2}\right)^2 - \left(\frac{u_1 - u_2}{2}\right)^2 \qquad \textbf{2.4}$$

so we can always express a product of two linear variables as a difference between two non-linear functions of linear variables. This is a special case of the fact that any quadratic function can be represented as a sum or difference of squares, and is therefore amenable to separable programming. And we could then handle the product of such a quadratic function with another variable in the same way; so in theory there is no limit to the class of functions that can be represented in this way. In practice of course a very involved representation would be cumbersome computationally.

So we can deal with a product term by introducing 2 groups of special variables. This involves 4 extra equations – 2 to represent the conditions that the sum of the special variables of each group must equal 1, and 2 to represent the value of the arguments of the non-linear functions in terms of the special variables. These correspond to Equations **2.1** and **2.2**. In practice we will generally not write down an equation corresponding to **2.3** explicitly, since we can substitute the left hand side of this equation for $f(z)$ wherever it occurs.

Since each product term involves 4 extra equations, we want to keep the number of product terms down as much as possible.

We therefore reformulate the problem defined under the heading 'Formulation' by introducing two new variables, P_1 and P_2, representing the proportion of the total sinter make sent to Ebbw Vale furnaces 1 and 2 respectively. We then replace the conditions **1.1** by

$$S_{ij} = P_j S_i \quad \text{for } i = 0, \ldots 8, j = 1, 2. \qquad \textbf{2.5}$$

The variables S_{i3} can be eliminated from the problem using **1.10**, since we can write

$$S_{i3} = S_i - S_{i1} - S_{i2}. \qquad \textbf{2.6}$$

Logically, we should have introduced the constraint

$$P_1 + P_2 \leqslant 1 \qquad \textbf{2.7}$$

to avoid the possibility of calling for a negative amount of sinter at Spencer Works. In practice we knew that this constraint would always be met, so we did not write it down explicitly.

Application of separable programming – defining the ranges

The next point to notice is that separable programming only deals with non-linear functions over definite ranges of values of their arguments. Even with such an elementary function as z^2 one must define realistic lower and upper limits for z – or else have either an unnecessarily inaccurate approximation or alternatively an unnecessarily large problem (with a large number of special variables in each group).

In practice, when dealing with product terms it seems best to define the quantities u_1 and u_2 for which one applies the identity **2.4** so that each covers the range $0 \leqslant u_1, u_2 \leqslant 1$.

To apply this philosophy we specified for the following quantities:

$S_{0\,min}$ = Minimum Contemplated Total Sinter Make
S_{0max} = Maximum Contemplated Total Sinter Make
P_{1max} = Maximum Contemplated Proportion of Sinter used at Ebbw Vale 1
P_{2max} = Maximum Contemplated Proportion of Sinter used at Ebbw Vale 2
I_{imin} = Minimum Percentage of Component i in Sinter ($i = 1 \ldots 8$)

$I_{i\max}$ = Maximum Percentage of Component i in Sinter ($i = 1 \ldots 8$)

So the minimum possible value of S_0 is $S_{0\min}$, and its range of possible value is S_{0R}, where

$$S_{0R} = S_{0\max} - S_{0\min}. \tag{2.8}$$

Similarly, the minimum possible value of S_i for $i = 1, \ldots, 8$ is $S_{i\min}$ where

$$S_{i\min} = 0{\cdot}01 \times I_{i\min} \times S_{0\min} \tag{2.9}$$

and its range of possible values is S_{iR}, where

$$S_{iR} = 0{\cdot}01 \times I_{i\max} \times S_{0\max} - S_{i\min}. \tag{2.10}$$

So we can rewrite **2.5** in the form

$$S_{ij} = S_{i\min} \times P_j + P_{j\max} \times S_{iR} \left(\frac{P_j}{P_{j\max}} . \frac{S_i - S_{i\min}}{S_{iR}} \right). \tag{2.11}$$

(In practice, different components were measured in different units to keep the numbers of comparable size – but this is a routine precaution in linear programming work that needs no special emphasis here.)

We have now succeeded in expressing our product terms in quantities

$$\frac{P_i}{P_{j\max}} \quad \text{and} \quad \frac{S_i - S_{i\min}}{S_{iR}}$$

that cover the interval 0 to 1. So if we again rewrite **2.11** in the form

$$S_{ij} = S_{i\min}.P_j + P_{j\max}.\, S_{iR}\,(z_1^2 - z_2^2), \tag{2.12}$$

where $$z_1 = \tfrac{1}{2}\left(\frac{P_j}{P_{j\max}} + \frac{S_i - S_{i\min}}{S_{iR}}\right) \tag{2.13}$$

and $$z_2 = \tfrac{1}{2}\left(\frac{P_j}{P_{j\max}} - \frac{S_i - S_{i\min}}{S_{iR}}\right), \tag{2.14}$$

we know that $0 \leqslant z_1 \leqslant 1$ and also $-\tfrac{1}{2} \leqslant z_2 \leqslant \tfrac{1}{2}$.

Defining the grid

Just one further decision has to be taken: how fine a grid of points do we need to represent our non-linear function? It is a straightforward matter to show that if we take a spacing of 0·2 units in z_1 and z_2 – giving us 6 points on each function, then the maximum possible error in $z_1^2 - z_2^2$ is 0·01. This corresponds to 1 per cent of its maximum value, and might have been adequate. But the actual error oscillates between plus and minus this maximum error, and we feared that such oscillation might increase the risk of finding an irrevelant local optimum to our problems. We, therefore, chose a spacing of 0·1 units, giving us 11 points on each function and a maximum possible error in $z_1^2 - z_2^2$ of 0·0025. As it turned out, this decision increased the computer time required to solve our problems significantly. The right solution in theory is to start with a coarse grid and to interpolate a finer one in the regions of greater interest. A programme to do this automatically is now being tested, but it was not available when we did this work originally; and it would have been a messy and time-consuming task to try to do this interpolation manually.

Computational experience

A set of related problems were solved in this way. They all involved 36 groups of special variables (making 396 such variables in all), together with 143 ordinary variables. The constraints numbered 168 in the first problem, with a few more added later. These constraints were made up as follows:

9 of the form **1.9** defining the values of S_i in terms of the inputs to the sinter plant

18 of the form **2.12** defining the values of the S_{ij} in terms of P_j and the special variables

36 of the form **2.1** defining the fact that the sum of the special variables in each group must equal 1

36 of the form **2.2** with the argument z replaced by its definition in terms of P_j and the S_i given by **2.13** and **2.14**

12 defining the composition of the products made at Spencer Works, with the effect of the sinter given by **2.6**

26 defining the composition of the products made at the 2 Ebbw Vale furnaces

31 miscellaneous constraints

After the coefficients of the first problem had been finally corrected, this took some 30 minutes to solve using the C-E-I-R. code LP/90 on the I.B.M. 7090 computer. Subsequent problems were solved starting from the most relevant existing solution to a previous problem, and took times that varied from 5 minutes up to about 30 minutes.

Implementation

Results

Work on this model started in the summer of 1963. At this time we did not know precisely which ores would be available for 1964 in what quantities nor what the exact prices would be. The first series of computer runs was carried out therefore without commercial constraints except for a few materials, and on the previous year's price list. Despite the limited usefulness of the results, it turned out a very good thing that these runs had been carried out, since the initial technical restrictions supplied to the computer were not completely accurate. Previous experience in applying linear programming to blast furnace burdening had shown that it is impossible to formulate these restrictions adequately first time off; indeed the restrictions are largely based on blast furnace management's experience, and this implies both that they may be difficult to express mathematically and that they may alter as further operating experience is gained.

The procedure for negotiating ore contracts with B.I.S.C. (Ore) is not rigid. In this case a list of requirements was submitted to B.I.S.C. (Ore) who examined it to determine whether the ores were available and suggested alternatives where they were not; agreement was eventually reached on a list of ore requirements satisfactory to both parties. During the negotiation period, B.I.S.C. (Ore) informed the steel companies of certain changes in ore prices as contracts with suppliers were placed.

Such a procedure creates a need to carry out several runs, as prices and commercial restrictions alter. Altogether about a dozen different sets of conditions had to be studied. However, not all of these required setting up as new Separable Programming problems on the I.B.M. 7090. After a time, sufficient experience was gained, from studying both a number of different solutions, and the ancillary information available on request from the LP/90 code (dual solutions, and cost and right-hand side ranging) to enable approximations to be made so that we could obtain new solutions from an ordinary LP and indeed, for some of the changes necessary almost by inspection. It must be stressed that we could not have done this without the earlier solutions to the complete model.

Discussion

Far too little is known about the details of the reactions within the furnace for burdening to be regarded as a purely economic and computational problem. Each burden produced by the computer needs to be vetted by experienced blast furnace management and the computer cannot replace their judgement. It therefore follows that the implementation of this project is not as cut-and-dried an affair as it may be in other contexts. Nevertheless some comments may be made here on the usefulness of the computer results to those responsible for ore purchase.

During the course of the computer runs, burdens were being produced both by ourselves and the works blast furnace staff. These burdens were initially widely divergent; by the end of the exercise when agreement was reached with B.I.S.C. (Ore) they were for all practical purposes identical. This was due to three factors: the addition of fresh restrictions to (and correction of errors in) the computer programme; the information being received from B.I.S.C. (Ore); and last but by no means least, the influence of the computer results on the views of the blast furnace people.

The standard cost for iron for 1964 has been considerably reduced from that for 1963. It is too early yet however to say how the burdens have worked in practice, and changes will inevitably be made from time to time. Even if the reduction in cost is fully realised it would not be correct to attribute this as a saving to the computer runs or to the project, since the whole exercise was carried out jointly with the blast furnace management, scientific staff and the accountants and in any case the possible savings were continually varying owing to changes in the ores available and their prices. However, the total bill for imported ore for one year runs into tens of millions of pounds, so that a very small percentage saving will amply justify the cost of quite lengthy computer calculations.

References

LAWRENCE, J. R., and FLOWERDEW, A. D. J. (1963), 'Economic models for production planning', *Operational Research Q.*, vol. 14, no. 1, pp. 11–29.

MILLER, C. E. (1963), 'The Simplex method for local separable programming', in R. L. Graves and P. Wolfe (eds.), *Recent Advances in Mathematical Programming*, McGraw-Hill, pp. 89–100.

VAJDA, S. (1960), *An Introduction to Linear Programming and the Theory of Games*, Methuen and Wiley.

8 H. Albach

Long-Range Planning in Open-Pit Mining

H. Albach, 'Long-range planning in open-pit mining', *Management Science*, vol. 13, 1967, pp. B549–68.

In open-pit mining of lignite, production plans have to be set up for a period of about twenty to thirty years. This is partly due to the close interaction of production plans and investment plans and partly due to legal requirements for open-pit mining in Germany. The major uncertainties encountered in setting up a production plan stem from the geological structure of the pit. The total deposit of lignite in the field as well as the stratification of the layers of waste material and lignite can be considered as stochastic variables. The production plan is formulated as a chance-constrained programming problem. The model requires maximization of a linear form subject to linear and non-linear constraints. In order to facilitate computation of the large-scale problems encountered in practical applications the original model is changed into a straightforward linear programming model. An iteration procedure is derived by which the solution to the original non-linear problem is found. The production plan is computed for different levels of acceptable risk. The results form a risk-profit-surface from which management has to pick the optimum-optimorum plan according to its risk-preference function.

Long-range planning in the firm is generally associated with planning for long-term investments in plant and equipment. These decisions inolve a high degree of uncertainty of future demand and technical progress. In fact, the longer the period over which the planning process extends, the more fundamental is this aspect of uncertainty. Long-range planning in the firm requires an explicit treatment of uncertainty. Quantification of uncertainties in investment problems involves a major problem of information. Another class of long-range planning problems in the firm deals with multi-period production schedules to meet uncertain demands during these periods.

This paper reports on work done in the field of yet another class of long-range planning problems. It deals with the determination of optimal multi-stage production plans with uncertainties on the side of production. It has a practical problem as its background, namely open-pit mining of lignite.

The Problem

In general, firms in the lignite industry operate several pits. The coal which is extracted from these pits is either shipped to briquet factories or to electricity plants. The dust content of lignite must be lower if it is used for molding briquets than if the lignite coal is used for heating steam generators, but these differences will be disregarded here. The problem can thus be considered as that of production of a single commodity for the demand of briquet and electricity plants. It is common for lignite mining companies to sign long-term delivery contracts with electricity plants. These contracts pertain to quantities as well as to prices and provide for high penalties if the delivery obligation is not met. These delivery contracts extend over a period of approximately twenty years. The demand for lignite by electricity generating firms can thus be considered given for each period up to the end of the planning period. Quantities of lignite produced in excess of these delivery obligations are shipped to briquet factories.

The major uncertainty in lignite mining lies therefore in the hazards of production. While the risk of a breakdown of one of the huge dredgers used for mining lignite today cannot be wholly neglected, the major uncertainty rests in the geological structure of the pit.

A lignite working has a certain horizontal dimension. The coal deposits form one layer in some places. In other places the surface sank faster during one period than during another, so that layers of sand of up to 120 feet in depth are found between the different strata of coal. At most four layers of coal are found in one pit. While in the old pits the first layer of coal was found several feet below the surface, in the deep pits which were opened after the war, huge layers of fertile soil, sands, and clay have to be removed.

Before a pit is opened, test bores are needed to ascertain the geological profile of the pit. These profiles are used to compute the total deposit of lignite in the pit. They also form the basis for the long-range production plans. The results of the test bores are therefore vital information for the planning process. The horizontal extension of a lignite pit is shown in Figure 1. Figure 1 shows the contours of the major coal deposit in the pit. The other layers may, of course, differ somewhat in shape. The line AA′ marks the dominant dislocation of the geological stratification

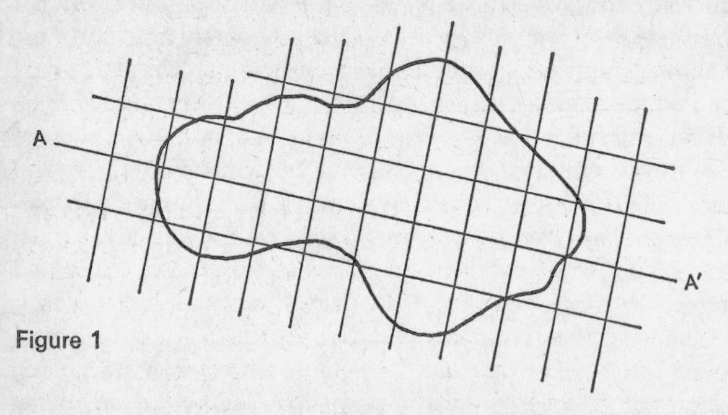

Figure 1

in the area. Parallel and vertical to the line AA′ some lines are drawn which form the grid of the test bores brought down by the firm's geologists to determine the quantity and quality of the coal deposit. The grid is in general 1500 by 1500 feet wide. In the neighborhood of major dislocations, the opening of the grid is brought down to 750 by 750 feet.

The results of the test bores are highly dependent on the boring method employed. Formerly dry boring was used most frequently. This method was very time-consuming and expensive. It took one week to drill 100 feet. In addition, the stratification of the ground could not be measured exactly, because in spite of lining the hole with pipes, material of higher layers used to fall down into the hole and was pulled out with the next boring cylinder. More recently, a rinsing procedure has come to be employed. By this procedure 100 feet can be drilled in one day. The disadvantage, however, is that it brings out the geological stratification of the ground even worse than the dry boring procedure. A major improvement of the boring results was achieved when an electrical method came to be used after the hole had been drilled. By means of a probe which is slowly dropped into the hole, electric current is sent out at one end. Since layers of gravel and clay are better conductors than the water in the hole, the current flows through these layers and back into the other end of the probe. By measuring the rheostat in the different layers, the stratification of the ground can be determined. However, the results achieved by this procedure may still be incorrect because (i) coal and clay have practically the same specific rheostat; and (ii) when strata are reached in greater depth which formerly were part of the ocean, the salt water in the hole becomes a better conductor than the surrounding layers of ground and the current no longer flows through these layers. While this last problem has not been overcome yet, the first obstacle to exact results was overcome by measuring the isotope K 40 which is found in clay but not in coal.

The profiles of the pit are then constructed by interpolating between the results of the test bores. Since formerly only parts of the total field were surveyed, interpolation in some places has to be carried out between results of different boring methods. These profiles consist of several vertical projections. Figure 2 gives a simplified example of such a projection taken at the line AA′ in Figure 1.

From this discussion it is obvious that two main factors of uncertainty result from the geological survey: the incorrect boring results and the interpolation between boring points. Interpolation leaves out smaller dislocations which range from six to twenty feet. The borderlines between layers of waste material and coal cannot be specified exactly in advance.

Trying to quantify these uncertainties is rather difficult.The only basis for quantification are the experiences gained in other open pits. The actual proportions of waste material and coal which can be compared with the data obtained by boring of the coal field are derived from the production reports. These reports contain many measuring errors. In wet weather more clay sticks to the sides of the carriages which are used to remove the waste material from the open pit than in dry weather with the result that the reports of waste material removal which do not take this factor into account overstate the removal quantities. Voluntarily slanted reports are not infrequent in this industry.

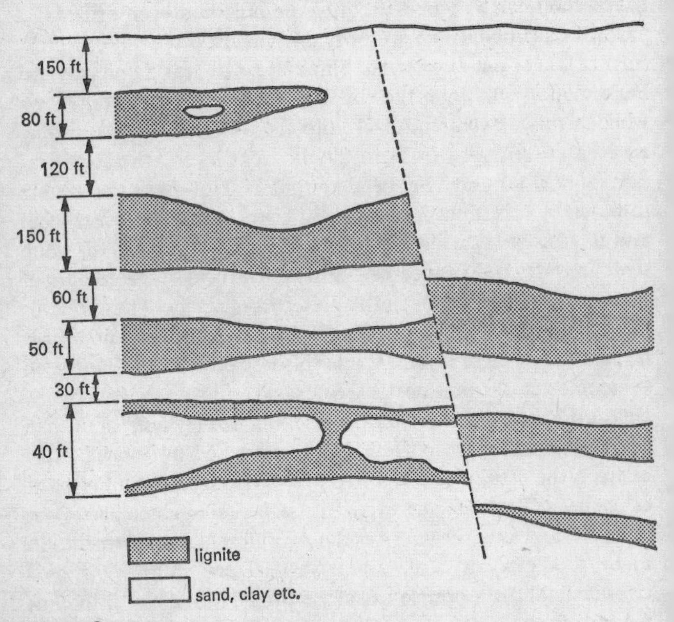

Figure 2

Past experience thus shows that the disparity between the actual geological circumstances in the pit and the planned structure of the pit on the basis of the boring measures is caused by a

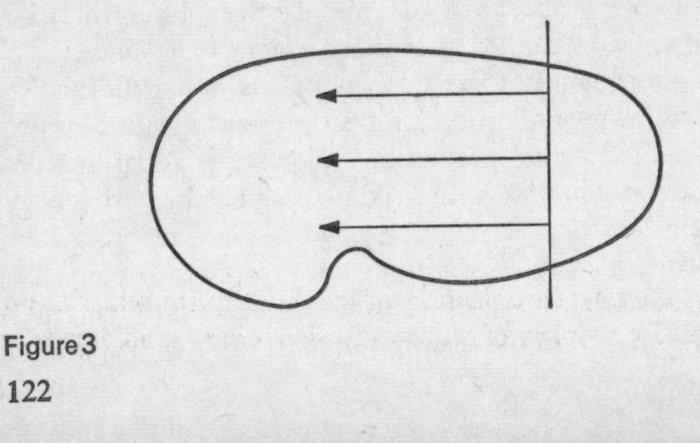

Figure 3

multitude of fairly independent factors. Differences between expected and actual states of the pit have varied between 1 and 20 per cent. On the average, however, actual circumstances have been fairly close to the expected values.

Two basic methods of mining lignite in open pits can be distinguished, namely, parallel dredging and circular dredging. Parallel dredging moves from a base line as shown in Figure 3, while circular dredging produces in rays around a center point as shown in Figure 4.

Production moves forward on different levels of the pit so that in general waste material is cleared away first, and in subsequent time-periods the coal below this waste material is mined. Figure 5 shows this production sequence in a vertical projection of a particular state of the exhaustion of the pit.

The first waste material taken off the coal layers has to be dumped on hills outside the pit. After a certain opening of the pit has been produced, all the remaining waste material can be dumped in the pit itself after the coal has been removed. The local boards of mining authorize a plan for the exhaustion of the deposit only if it guarantees a minimum waste material disposal outside the open pit.

In the low open pits the proportion of waste material to coal amounted to about 1:2 so that waste material removal did not constitute a serious problem. In the deep pits which are being mined now, the proportions have shifted to about 4:1 so that dredging and transportation of waste material takes up a major part of the capacity installed. While in the low pits removal of waste material and production of coal could be considered fairly independent of each other, production in the deep pits requires

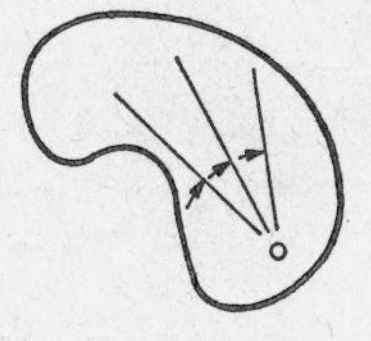

Figure 4

a fairly strict synchronization of coal and waste material production. The problem of synchronization is even more important

since the various layers of the large mass of waste material on top of the first stratum of coal have to be reproduced in the same order when the pit is refilled and recultivated after the removal of the

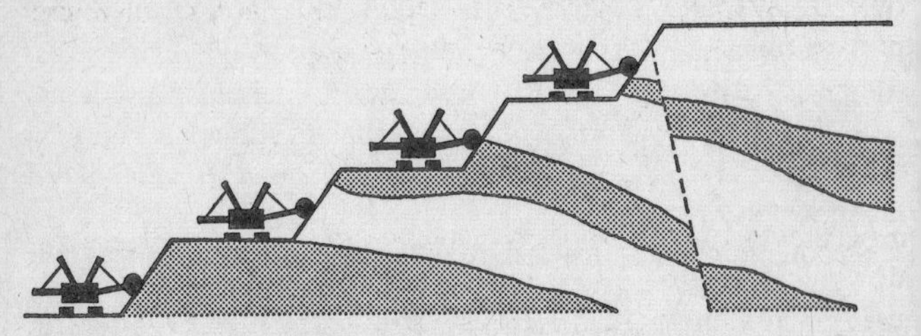

Figure 5

coal. However, it is not possible to say that removal of waste material and coal production are strictly dependent on each other. A certain amount of freedom to remove different quantities of waste material for a given quantity of coal produced is still prevalent even in the deep pits.

The Model

Long-range production planning under certainty

We shall now formulate the model which is designed to compute the optimal long-range mining plan for lignite. Throughout, the assumption will be made that the stratification of waste material and coal is horizontal. As a starting point the model of long-range coal production is formulated on the assumption that the geological measurements coincide with reality.

First, we shall assume that there will always be synchronized production of waste material and coal. We then get the following model: If k_{it}^{C} is the cost of producing one ton of coal from pit i in time-period t and k_{it}^{R} is the corresponding cost of removing one ton of waste material, then the total cost k_{it} of one ton of coal from pit i in time-period t is given by

$$k_{it} = k_{it}^{C} + \frac{a_i}{k_{it}^{R}} \qquad \mathbf{1}$$

where $1/a_i$ is the tons of waste per ton of coal at pit i. Since strata in the pit are horizontal and production is synchronized, a_i is a constant and independent of time. The expected price per ton of coal in period t is p_t. Then gross profit per ton of coal from pit i in period t is the difference $p_t - k_{it}$.

The expected profit of the operation of n pits ($i = 1, \ldots, n$) over H time-periods that are included in the long-range planning process ($t = 1, \ldots, H$) can then be given by

$$E = \sum_{t=1}^{H} \sum_{i=1}^{n} (p_t - k_{it}) X_{it} q^{-t} \Rightarrow \max. \qquad 2$$

q^{-t} is the discount factor used to discount the net gain of period t down to the present [$q^{-t} = 1/(1 + r)^t$ with r the interest rate used for discounting]. X_{it} is the amount of coal produced from pit i in period t.

The following constraints are to be observed. The total amount of coal extracted from pit i up to the planning horizon must be no greater than the deposit S_i (in tons of coal) in pit i

$$\sum_{t=1}^{H} X_{it} \leqslant S_i \quad \text{all } i. \qquad 3$$

Dredger capacity K_{it} installed in pit i during period t sets a limit to the amounts producible from the pit

$$(1 + 1/a_i) X_{it} \leqslant K_{it} \quad \text{all } i, t. \qquad 4$$

It may be added that dredgers will never be moved from one pit to another in the course of one year due to the fact that they are too huge to be moved without dismantling them to a certain extent.

The lignite production from all pits is required to be no less than the deliveries contracted for period t. Delivery requirements are denoted by the symbol N_t.

$$\sum_{i=1}^{n} X_{it} \geqslant N_t \quad \text{all } t. \qquad 5$$

Surface water management in an open pit requires that the surface of the pit should not change too drastically from one year to the other. Transportation installations, i.e. carriages for waste and conveyor belts for coal, impose further constraints on the extent of changes of the level of production. Also, it is the declared policy of the company to stabilize employment. These conditions can be combined in a production-smoothing constraint

$$\mu_{ui} X_{it-1} \leqslant X_{it} \leqslant \mu_i^0 X_{it-1} \quad \text{all } t \qquad 6$$

with $\mu_{ut} - \mu_i^0$ expressing the range of coal production from pit i in period t as a proportion of the previous years' coal production. The range factors may or may not be symmetrical around the previous year's production figures. In the study the ranges were set at ± 20 per cent. The non-negativity constraints have to be observed

$$X_{it} \geqslant 0 \quad \text{all } i, t. \qquad 7$$

The model has to be changed only slightly if it is assumed that waste material and lignite can be produced independently. The expected profit is then

$$E = \sum_{t=1}^{H} \sum_{i=1}^{n} (p_t - k_{it}^C) X_{it} q^{-t} - \sum_{t=1}^{H} \sum_{i=1}^{n} k_{it-1}^R Y_{it-1} q^{-t} \Rightarrow \max \qquad 8$$

to be maximized subject to

$$\sum_{t=1}^{H} X_{it} \leqslant S_i \quad \text{all } i \qquad 9$$

$$X_{it} + Y_{it} \leqslant K_{it} \quad \text{all } i, t \qquad 10$$

$$\sum_{i=1}^{n} X_{it} \leqslant N_t \quad \text{all } t \qquad 11$$

$$\sum_{t=1}^{r} X_{it} - a_i \sum_{t=1}^{r} Y_{it-1} \leqslant 0 \quad \text{all } i, \quad (r = 1, \ldots, H) \qquad 12$$

$$\mu_{ui} X_{it-1} \leqslant X_{it} \leqslant \mu_i^0 X_{it-1} \quad \text{all } i, t \qquad (7a)$$

$$\mu'_{ui} Y_{it-1} \leqslant Y_{it} \leqslant \mu'^0_i Y_{it-1} \quad \text{all } i, t. \qquad (7b)$$

12 states the condition of independent coal and waste production. Waste material has to be removed some time, at least one period before coal production. However, as long as waste is removed well in advance it is not necessary to determine the period of waste removal in advance. In practice, the situation in which the equality sign holds is called 'production with run-up dredgers'. The technical people tend to avoid this situation for obvious reasons. The financial people on the other hand do not favor waste removal to be too far ahead of coal production because this ties up capital and is disadvantageous for tax purposes.

It is on principle necessary to differentiate between production smoothing constraints for coal **7a** and waste material **7b** because some of the conditions combined in these constraints operate differently for waste and for coal. This is particularly true with regard to the transportation and employment conditions. However, in this case the ranges could be set at ± 20 per cent for coal as well as for waste material.

To allow for capacity constraints on the amount of outside waste material dumping and on inside dumping during the period of opening the pit up it would be necessary to differentiate between waste material dumped outside and inside the pit. It is assumed here that these constraints are not binding in any period so that they can be neglected.

The models based on expected values only are straightforward linear programming problems. However, working out production plans on the basis of expected values is equivalent to accepting a probability level of 50 per cent of violating the constraints. This is, of course, not acceptable in general.

Long-range production planning under uncertainty

We will therefore introduce uncertainty explicitly into the model. The deposit S_i is a stochastic variable which is assumed to be normally distributed with mean $\bar{S}_i$ and variance $V[S_i]$. The ratio a_i is also a stochastic variable which can be assumed to be normally distributed with mean $\bar{a}_i$ and variance $V[a_i]$. The assumption of the normal distribution is not only mathematically convenient but also appropriate in view of the fact that all the factors causing the uncertainty with regard to the expected values seem to be fairly independent of each other and additive in character.

In formulating the model the basic idea of chance-constrained programming is used to incorporate these uncertainty components into the model. The general formulation of chance-constrained programming may be given as in Charnes and Cooper (1963; p. 263 in this volume):

Maximize $$E = E(X, Y; p, k) \qquad 13$$

subject to $$P(AX + BY \leqslant c) \geqslant \pi \qquad 14$$

and $$X, Y \geqslant 0 \qquad 15$$

where $$0 \leqslant \pi_j \leqslant 1. \qquad 16$$

P stands for probability. A and B are coefficient matrices, c is a requirements vector, and π is a vector of probabilities π_j ($j = 1, \ldots, m$) which mean that in a fraction π_j out of one cases this jth constraint has to be fulfilled, or vice versa: in only the $(1 - \pi_j)$th fraction of all the cases may the jth constraint be violated. Of course, these probability levels can in some cases not be specified in advance but depend on the problem itself. In the case reported on here, the probability level could be derived from the penalty for not meeting the delivery requirements specified by the long-term supply contract. Nevertheless it proved most interesting to examine the sensitivity of the solution to variations of the various probability levels.

The probabilistic constraints used in deriving the long-range production schedule will first be developed by taking as an example a single coal pit. It is obvious that the delivery requirements of all periods prior to the planning horizon H will be met if the requirement in the Hth period can be fulfilled, considering the uncertainty of the coal deposit only. Therefore the specific formulation of **14** for this type of uncertainty becomes

$$P(X_H \geqslant N_H) \geqslant \pi. \qquad 17$$

Now since $$X_H \leqslant S - \sum_{t=1}^{H-1} X_t \qquad 18$$

by definition, **17** becomes

$$P\left(S - \sum_{t=1}^{H-1} X_t \geqslant N_H\right) \geqslant \pi. \qquad 19$$

Now from the laws of the normal distribution

$$P(\bar{S} - \lambda\sigma \leqslant S) = 1 - \phi(\lambda) = \pi \qquad 20$$

with $\lambda(\pi)$ being the normal deviate and σ the standard deviation of S. Therefore **19** will be fulfilled with probability π if in

$$\sum_{t=1}^{H-1} X_t \leqslant \bar{S} - \lambda(\pi)\sigma - N_H \qquad \textbf{21}$$

the normal deviate is set at the level corresponding to π. Since λ is tabulated for any given π, **21** is a linear inequality for given π. **21** then is the certainty constraint which replaces the stochastic constraint **17**.

We now take the second source of uncertainty into consideration. Waste material has to be removed at least one year ahead of coal production. This means that meeting the delivery requirements is jeopardized if not enough coal is ready for removal at the beginning of the production period. Waste removal must therefore meet the probabilistic constraints

$$P\left(a \sum_{t=1}^{\tau} Y_{t-1} - \sum_{t=1}^{\tau-1} X_t \geqslant X_\tau \right) \geqslant \pi \quad \text{all } \tau. \qquad \textbf{22}$$

The left-hand side of the inequality in the parentheses stands for the amount of coal ready for removal at the beginning of period τ. X_τ is the coal production in period τ. **22** is equivalent to

$$P\left(\sum_{t=1}^{\tau} X_t - a \sum_{t=1}^{\tau} Y_{t-1} \leqslant 0 \right) \geqslant \pi \quad \text{all } \tau. \qquad \textbf{23}$$

Now a is a normally distributed variable with mean $\bar{a}$ and variance $V(a)$. Therefore we have from the normal distribution

$$P\left[a \sum_{t=1}^{\tau} Y_{t-1} - \lambda(\pi)\, \sigma(a) \left(\sum_{t=1}^{\tau} Y_{t-1}^2 \right)^{\frac{1}{2}} \leqslant a \sum_{t=1}^{\tau} Y_{t-1} \right] \geqslant \pi \qquad \textbf{24}$$

for all τ with $\sigma(a)$ the standard deviation of a assumed constant for all periods τ. From **23** and **24** we derive

$$\sum_{t=1}^{\tau} X_t - \bar{a} \sum_{t=1}^{\tau} Y_{t-1} + \lambda(\pi)\, \sigma(a) \left(\sum_{t=1}^{\tau} Y_{t-1}^2 \right)^{\frac{1}{2}} \leqslant 0. \qquad \textbf{25}$$

25 is the certainty equivalent to the stochastic constraint **22**. From all practical evidence it has to be assumed that the stochastic variables a and S are independent of each other. This is at any rate the assumption underlying the models.

To allow for uncertainty, the model of multi-period planning for n open pits and synchronized production given by **1** through **7** requires a new formulation of the capacity and the demand constraints only. It is seen that the capacity constraint **4** is replaced by

$$\left(1 + \frac{1}{\bar{a}_i - \lambda(\pi)\sigma(a_i)}\right) X_{it} \leqslant K_{it} \quad \text{all } i, t \qquad \mathbf{26}$$

and that the demand conditions **5** and **3** are now given by

$$\sum_{i=1}^{n} X_{it} \geqslant N_t \quad \text{all } t \qquad \mathbf{27}$$

$$\sum_{t=1}^{H} \sum_{i=1}^{n} X_{it} \leqslant \sum_{i=1}^{n} \bar{S}_i - \lambda(\pi)\left(\sum_{i=1}^{n} V[S_i]\right)^{\frac{1}{2}}. \qquad \mathbf{28}$$

Similarly in the case of independence between removing waste material and producing lignite in n open pits the following constraints allow for uncertainty. The inequalities **27** and **28** express the uncertainty of the coal deposit in this model, too, and replace **11** and **9** respectively. Allowance for uncertainty of removal conditions is made by

$$\sum_{t=1}^{\tau} X_{it} - \bar{a}_i \sum_{t=1}^{\tau} Y_{it-1} + \lambda(\pi)\,\sigma(a_i)\left(\sum_{t=1}^{\tau} Y_{it-1}^2\right)^{\frac{1}{2}} \leqslant 0 \qquad \mathbf{29}$$

for all i, r. **29** is substituted for **12** above. While the model with synchronized production is again a straightforward linear programming problem which can be solved parametrically for different values of $\lambda(\pi)$, the model for independent production is a problem of maximization of a linear function subject to linear and non-linear constraints. Some comments on the procedure used for numerical solution are therefore in order.

Computational Aspects

In order to solve the model with the constraints **29** for a firm which operates several pits and plans its activities over a period of about twenty years, it seems highly desirable to use existing LP-programs. The solution procedure was developed with this aim in mind. It is based on two conditions. The solution space is convex, and at least one linear constraint is binding in the optimal solution. A proof that the region of all vectors satisfying non-linear constraints of the type **29** is convex has been given by van de Panne and Popp (1963, p. 405). The second condition limits the generality of the procedure but this has not been found a disadvantage in application because the number of non-trivial linear constraints is great indeed. First the basic idea for solving the problem is demonstrated by a single non-linear constraint. Then the idea is extended to the many non-linear constraints of **29** encountered in the study.

Suppose we have the specific form of **29**

$$X_1 + X_2 \leqslant F = Y_0 + Y_1 - \lambda(\pi)\sigma(a)\,(Y_0^2 + Y_1^2)^{\frac{1}{2}}. \qquad \mathbf{30}$$

We now linearize **30** by adding the term $2Y_0Y_1$ to the sum of squares in the root of **30** and get

$$X_1 + X_2 \leqslant F^* = Y_0 + Y_1 - \lambda(\pi^*)\sigma(a)\,(Y_0 + Y_1). \qquad \mathbf{31}$$

31 requires more waste material to be removed for a given withdrawal of coal than would be necessary at the given risk preference. Or, expressed the other way around: by linearizing the non-linear constraint in this way, we narrow down the solution space for the X's. This is shown in Figure 6. A set of non-linear isoquants convex to the origin can be obtained from **30** for different values of F whereas from **31** we get a set of linear isoquants for different values of F^*. Let us suppose the value of π in the actual constraint **30** is predetermined by the company at π^0. Instead of **30** we use **31** in the computation and start with a value of π_1^* equal to π^0. We then get values of Y_{01}^*, Y_{11}^* as solutions of the linear program. We insert these solutions in **30** and solve for π to get, say, π_1. The relation $\pi_1 > \pi^0$ will hold. We then choose π_2^*, $\pi_2^* < \pi_1^*$. We then solve a new linear programming problem with **31** as the constraint used in the computation. Inserting the solution $Y_{02}^* Y_{12}^*$ in **30** and solving for π we get π_2 with $\pi_2 < \pi_1$ and, say, $\pi_2 < \pi^0$. Then π^* is set at π_3^* with $\pi_1^* > \pi_3^* > \pi_2^*$. Again the solution vector is used to solve for π in **30**. The iteration terminates if $\pi_3 = \pi^0$. This is shown graphically in Figure 7.

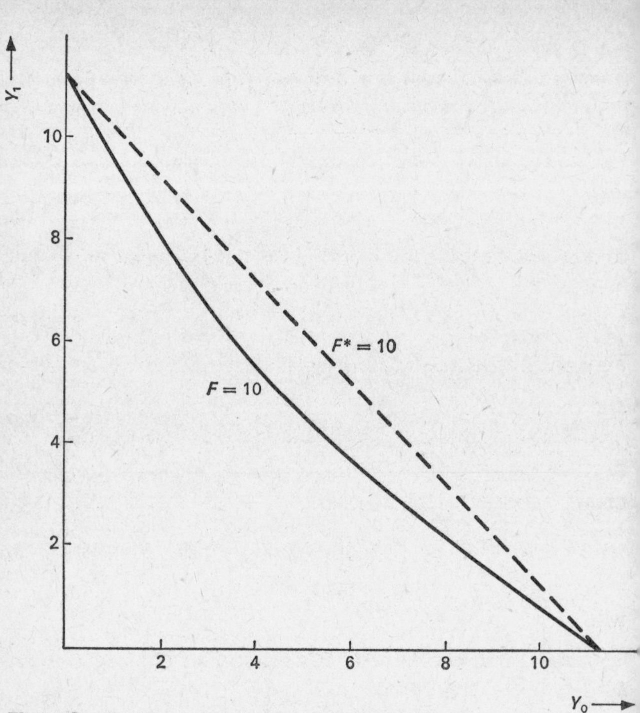

Figure 6

This basic idea is now applied to **29**. The safety margin

$$\lambda(\pi)\sigma[a_i]\left(\sum_{t=l}^{\tau} Y_{i\,t-1}^2\right)^{\frac{1}{2}}$$

is overstated by the linear safety margin $\lambda(\pi)\sigma(a_i)\sum_{t=l}^{\tau} Y_{it-1}$ by a fraction, say ρ, so that

$$\lambda(\pi)\sigma(a_i)\left(\sum_{t=1}^{\tau} Y_{i\,t-1}^2\right)^{1/2} = [\lambda(\pi) - \rho]\sigma(a_i)\sum_{t=1}^{\tau} Y_{it-1}. \qquad \mathbf{32}$$

$\lambda(\pi) - \rho$ is the deviate in the linearized form which corresponds to the prescribed normal deviate $\lambda(\pi)$ in the non-linear constraint. From **32** we get

$$\rho = \lambda(\pi) \left[1 - \frac{\left(\sum_{t=1}^{\tau} Y_{it-1}^2 \right)^{\frac{1}{2}}}{\sum_{t=1}^{\tau} Y_{it-1}} \right]$$

and by setting the second term in the brackets equal to $D_{i\tau}$

$$\rho = \lambda(\pi) [1 - D_{i\tau}] \tag{34}$$

and inserting in **32** and **29** we get

$$\sum_{t=1}^{\tau} X_{it} - [\bar{a}_i - \lambda(\pi)\sigma(a_i)D_{it}] \sum_{t=1}^{\tau} Y_{it-1} \leqslant 0 \quad \text{all } i, \tau \tag{35}$$

or

$$\sum_{t=1}^{\tau} X_{it} - \varnothing_{ri} \sum_{t=1}^{\tau} Y_{it-1} \leqslant 0 \quad \text{all } i, \tau \tag{36}$$

with

$$\varnothing_{\tau i} = \bar{a}_i - \lambda(\pi)\sigma(a_i)D_{i\tau}\,. \tag{37}$$

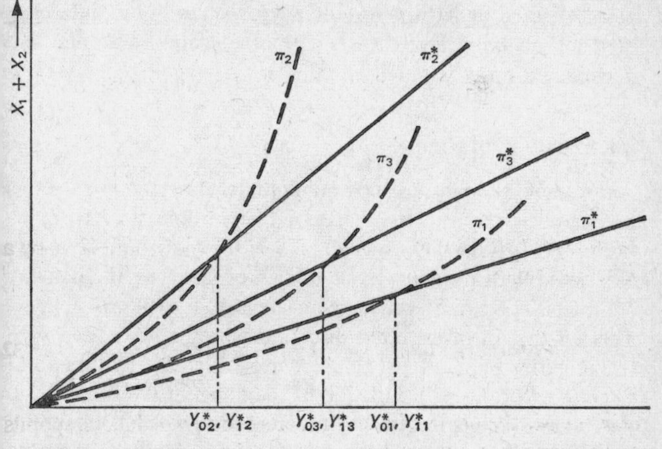

Figure 7

36 is a linear expression if the D_{ir} are given. The D_i depend on the solution. If the D_{it} are assumed given and the solution is obtained in terms of these D_{ir}^0, new values of D_{ir}^1 can be computed from this solution. These values are used for the next iteration and so forth until $D_{ir}^m = D_{ir}^{m+1}$.

Two basic situations may now be distinguished: 1) none of the non-linear constraints is binding, and 2) at least one of the non-linear constraints is binding. In the first case the values of D_{ir}^m correspond to a λ_{ir}^m (π_{ir}^m) that is larger than the prescribed level of $\lambda(\pi^0)$. At this level, however, the linearized form ceases to be a binding constraint. Then from the convexity of the region the solution obtained in the mth iteration is the optimal solution. The optimal solution is determined by the linear constraints only. The non-linear constraints may be dropped, and the resulting linear program yields the same solution as the iterative procedure. In the second case the value of D_{ir}^m corresponds to a $\lambda_{ir}^m(\pi_{ir}^m)$ which is equal to $\lambda(\pi^0)$. This solution is the optimal solution if in the optimal solution at least one linear constraint of the original non-linear program is binding. The solution is not (with the exception of the case of degeneracy) the optimal solution if this condition does not hold. Convergence of this procedure has been found to be quite rapid. Convergence may be accelerated by letting the step range vary with the difference in the probability levels $\pi^m - \pi^{*m}$ of the non-linear and the linearized form.

Interpretation of Results

An example of long-range planning under uncertainty with this procedure is given in the Appendix. The results may be interpreted here, however. Figure 8 shows total discounted profits associated with different levels of risk of violating the delivery contracts. These risk levels are here labelled 'deposit probability'. There is a gain in expected discounted profits associated with relaxing the required probability level. Figure 9 shows total discounted profits as a function of the different levels of risk in the removal constraints. By the term 'removal probability' we express the predetermined probability of not having removed waste material in prior periods that is sufficient to guarantee the

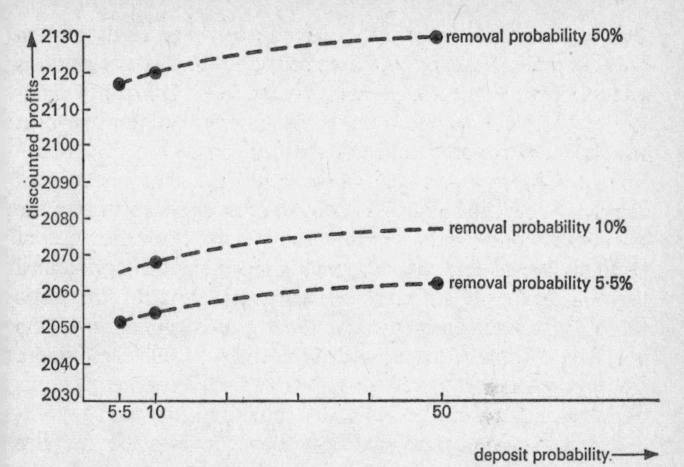

Figure 8

coal production necessary to satisfy the demand contracted for the current period. It is interesting to note that total discounted profits are more sensitive to variations in the probability level of the removal constraints than to variations in the probability

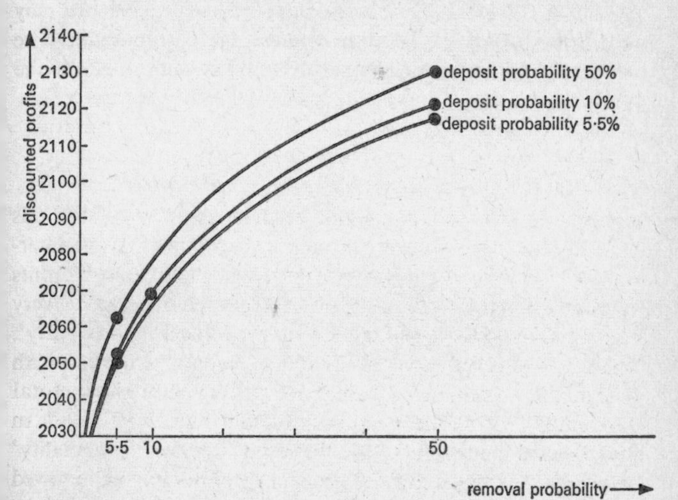

Figure 9

level of the deposit constraints. This is, however, easily understood because the removal constraints exert a greater influence on production in the first periods than do the deposit constraints. Figure 10 shows the combined effect of the two risks involved. The various risk-profit alternatives form a risk-profit surface. It is, of course, for management to pick the risk-profit-combination in line with management's risk-preference-function as the optimum optimorum long-range production plan of the firm. Consideration of the penalties payable upon violating the contracts will naturally set an upper limit for the risks acceptable to management. The lower bound on risk aversion is largely determined by competitive forces. Within these limits factors of personality are generally responsible for the choice of management among various risk-profit-alternatives.

The implications of the models developed here for the planning process of the company are the following. The actual planning process in the firm covers seven large pits out of which two were run with somewhat changed figures in the example. Ideas about opening a new pit within the next twenty years are being discussed within the firm and with state authorities. However, these considerations were judged too vague to be incorporated in the long-range plans. Long-range plans of the type developed here are used for two purposes. They are used as a basis for co-ordinating the supply of lignite with the supply of other sources of electric power within the long-range plans of the utility company that controls the lignite company. Also, investment plans of the utility company have to be co-ordinated with the investment plans of the lignite company. The long-range plans are used for this purpose as well. In this phase of the planning process delivery requirements and installed capacity are treated as parameters that are changed to provide alternative plans for the co-ordinating process. The second purpose of these long-range plans takes delivery requirements and installed capacity as given. The long-range production plan is then used as a basis for long-range financial and personnel planning as well as for co-ordination with the mining authorities. The law requires the firm to submit long-range production plans as well as short-run operating plans to the mining authorities for approval. The long-range plans have to be submitted because devastation of the surface

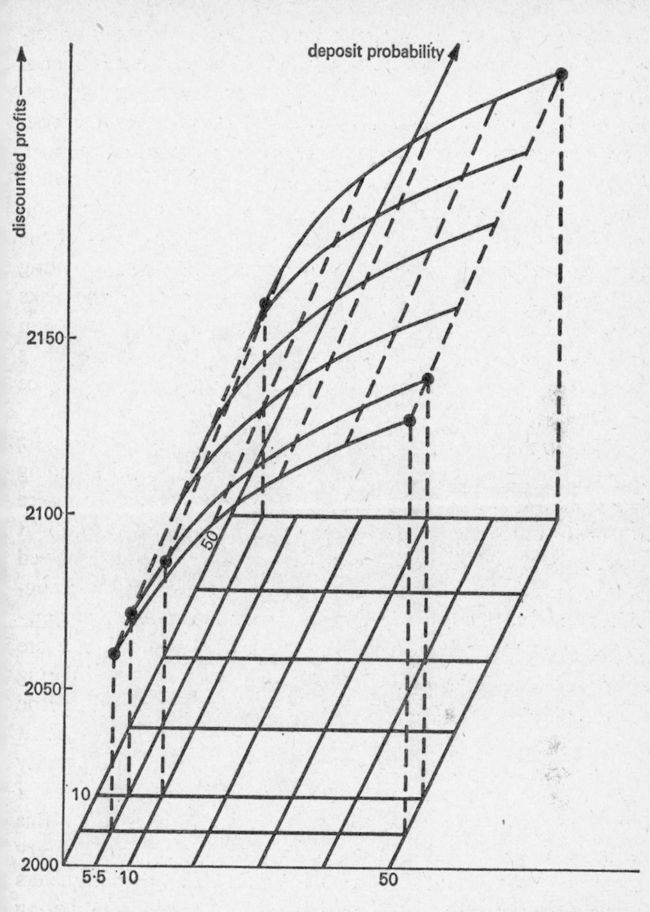

Figure 10

in the densely populated area requires dislocation of villages, railroads, highways, and streams, and has therefore to be co-ordinated with the state planning board. The operating plans are used by the authorities to impose safety regulations of various specific kinds. The operating plans have to be integrated with the long-range plans submitted by the company. Since formerly uncertainties in the geological factors influencing production have been widely neglected in the planning process, changes of the long-range plans as well as the operating plans submitted to the authorities were frequent. It is the aim of management to cut down the number and the extent of these changes. Therefore, the risk aversion with regard to the removal constraint is higher than the risk aversion with regard to the deposit constraint. The deposit constraints show their impact during the later phase of the plan, and adjustments may be made in the meantime. The removal constraints have an immediate impact on operations. A high risk aversion in the removal constraint leads to high investment in dredgers and to high 'stocks of waste material' which management would like to avoid because they tie up capital and are not tax-deductible. Also these stocks of waste material are a constant source of discussion with mining authorities if they have to be dumped outside the pits where they form huge hills.

The risk-profit-alternatives supplied by these models are therefore considered an appropriate basis for choice by management because they help to strike a balance between:

1. Lost profits due to inability to meet demand and high cost of capital tied up in capacity and stocks of waste material.

2. Immediate profits due to selling to briquet factories and penalties for not being able to meet delivery requirements in later periods (future profits on the briquet market in the light of today are not only lower than immediate profits due to time preference but also because the demand for briquets is diminishing).

3. High cost of co-ordination with authorities due to request for approval of large quantities of outside dumping and high cost of operating difficulties within the pits if requirements for outside dumping are low.

Appendix

Computation of the optimal production schedule

A firm operates two open pits and wants to determine the optimal production schedule over a period from 1961 to 1980. The long-term delivery contracts provide for a price per ton of coal of 10 marks, and deliveries have to be made as shown in Table 1.

The capacity expansion envisaged to operate the two pits is planned as shown in Table 2.

The expected profit margins per ton of coal and the expected costs of removal of one ton of waste material are given in Table 3.

Table 1
Delivery Requirements of Coal (millions of tons)

1962	27·4	1967	24·5	1972	31·7	1977	27·2
1963	29·2	1968	26·4	1973	31·7	1978	27·2
1964	29·2	1969	29·0	1974	31·7	1979	25·4
1965	27·4	1970	31·7	1975	31·7	1980	25·4
1966	24·5	1971	31·7	1976	31·7	Total	554·7

Table 2
Installed Capacity (millions of tons)

Year	*Pit 1*	*Pit 2*	*Year*	*Pit 1*	*Pit 2*
1961	60·00	65·00	1971	70·00	94·83
1962	70·00	90·00	1972	70·00	94·83
1963	70·00	90·00	1973	70·00	94·83
1964	70·00	90·00	1974	70·00	94·83
1965	70·00	90·00	1975	70·00	94·83
1966	70·00	90·00	1976	70·00	81·43
1967	70·00	90·00	1977	70·00	76·03
1968	70·00	90·00	1978	70·00	70·67
1969	70·00	91·59	1979	70·00	68·51
1970	70·00	94·83	1980	11·25	19·68

Ten per cent is used as the discount rate throughout. This rate was set by management and consists of a base rate of 7 per cent and a risk premium of 3 per cent. The theoretical validity of such a rate will not be argued here. The ratio of coal to waste material is 0·243 for Pit 1 and 0·403 for Pit 2. The coefficients of variation of these ratios are estimated to be 7 per cent in each lignite field. The total deposits of coal in the fields are estimated at 220 and 430 million tons respectively. The coefficient of variation of the total deposit in each pit is also 7 per cent. The optimal production plan for coal is given in Tables 4 and 5. The optimal plan for the removal of waste material is given in Tables 6 and 7.

Table 3
Discounted Expected Profit Margins Per Ton of Coal and Discounted Expected Costs Per Ton of Waste Material (Deutschmarks)

Year	*Pit 1*		*Pit 2*		Year	*Pit 1*		*Pit 2*	
	Coal	*Waste*	*Coal*	*Waste*		*Coal*	*Waste*	*Coal*	*Waste*
1961	0·00	1·28	0·00	1·20	1971	5·38	0·76	5·43	0·71
1962	8·30	1·22	8·21	1·31	1972	5·12	0·73	5·17	0·68
1963	7·87	1·20	8·11	0·96	1973	4·87	0·70	4·92	0·65
1964	7·57	1·07	7·76	0·88	1974	4·64	0·66	4·64	0·74
1965	7·20	1·03	7·35	0·88	1975	4·41	0·64	4·34	0·71
1966	6·88	0·97	6·98	0·86	1976	4·20	0·61	4·24	0·57
1967	6·53	0·93	6·62	0·84	1977	4·02	0·56	4·03	0·55
1968	6·23	0·88	6·31	0·80	1978	3·82	0·54	3·84	0·52
1969	5·92	0·85	6·00	0·77	1979	3·64	0·52	3·60	0·50
1970	5·64	0·81	5·71	0·74	1980	3·21	0·74	3·48	0·48

Tables 4 through 7 show the effects of risk aversion on coal production and waste removal plans. An increase in the probability level in the removal constraints leads to an increase in waste removal, particularly during the first part of the total planning period. Since capacity is fixed, coal production decreases accordingly. The situation is reversed during the later part of the planning period. The effect of risk aversion in the deposit constraint can also be seen from these tables and especially from the figures for Pit 1. During the period from 1962 to 1968 time preference is stronger than risk aversion in the deposit constraint: installed capacity limits production from Pit 1 until 1967. Then from 1969 to 1973 coal production is held off in order to be able to meet delivery requirements which are binding from 1974 on until the end of the total planning period.

The profits associated with these production plans are given in Table 8. Table 8 also shows the most interesting shadow prices obtained. The shadow price for total deposit is related to the constraint

$$\sum_{t=1961}^{1980} \sum_{i=1}^{2} X_{it} \leqslant 650$$

of the model, with 650 being the expected total of the deposits in the two pits. The shadow price for removal is derived from the two removal constraints

$$X_{1,1962} - 0{\cdot}243\ Y_{1,1961} + \lambda\,(\pi)\,0{\cdot}01701\,Y_{1,1961} \leqslant 0$$
$$X_{2,1962} - 0{\cdot}403\ Y_{2,1961} + \lambda\,(\pi)\,0{\cdot}02821\,Y_{2,1961} \leqslant 0.$$

Thirty-eight non-linear constraints were in the program. Convergence was quite rapid. After the third iteration the solutions were identical up to the fifth decimal.

Table 4
Coal Production Plan (millions of tons, deposit probabilities 50 per cent for each pit)

Year	*Pit 1, Removal probability*			*Pit 2, Removal probability*		
	50%	*10%*	*5·5%*	*50%*	*10%*	*5·5%*
1962	14·5800	13·2720	12·9012	26·1950	23·8453	23·1790
1963	13·4671	13·2930	13·2364	25·7134	25·6565	25·6198
1964	13·7375	13·3962	13·3013	25·9075	25·2040	25·0140
1965	13·6718	13·4308	13·3618	25·8293	25·4867	25·3860
1966	13·6878	13·4593	13·3929	25·8609	25·4528	25·3386
1967	13·6839	13·4798	13·4216	25·8481	25·5178	25·4197
1968	12·8347	13·4952	13·4422	25·8532	25·5293	25·4437
1969	10·2677	12·4403	13·1539	25·8512	25·5617	25·4710
1970	8·2142	9·9850	10·5720	26·4928	26·1921	26·1115
1971	6·5714	8·0136	8·4855	27·5399	27·2323	27·1475
1972	5·2571	6·4189	6·8064	26·6058	26·8577	26·7807
1973	4·2057	5·1351	5·4451	27·4943	27·0147	26·9380
1974	9·7945	10·0257	9·8369	21·9955	21·6743	21·8632
1975	14·1036	14·3606	14·2095	17·5964	17·3394	17·4905
1976	14·1310	12·1891	11·4153	17·5690	19·5109	20·2847
1977	13·5762	9·7705	9·1693	13·6238	17·4295	18·0308
1978	13·7110	14·4463	14·5277	13·4890	12·7537	12·6722
1979	13·6782	13·3387	13·2709	11·7218	12·0613	12·1291
1908	6·5868	5·7200	5·7200	18·8132	19·6800	19·6800
Total	*215·6702*	*215·6701*	*215·6699*	*430·0000*	*430·0000*	*430·0000*

Table 5
Coal Production Plan (millions of tons, removal probabilities 5·5 per cent for each pit)

Year	*Pit 1, Deposit probability*			*Pit 2, Deposit probability*		
	50%	*10%*	*5·5%*	*50%*	*10%*	*5·5%*
1962	12·9012	12·9012	12·9012	23·1790	23·1790	23·1790
1963	13·2364	13·2364	13·2364	25·6198	25·6198	25·6198
1964	13·3013	13·3013	13·3013	25·0140	15·0140	25·0140
1965	13·3618	13·3618	13·3618	25·3860	25·3860	25·3860
1966	13·3929	13·3929	13·3929	25·3386	25·3386	25·3386
1967	13·4216	13·4216	13·4216	25·4197	25·4197	25·4197
1968	13·4422	13·4422	13·4422	25·4437	25·4437	25·4437
1969	13·4606	13·1539	12·7875	25·4710	25·4710	25·4710
1970	11·9195	10·5720	10·2806	26·1115	26·1115	26·1115
1971	9·5806	8·4855	8·2515	27·1475	27·1475	27·1475
1972	7·6846	6·8064	6·6186	26·7807	26·7807	26·7807
1973	6·1477	5·4451	5·2949	26·9380	26·9380	26·9380
1974	9·8319	9·8369	9·8408	21·8681	21·8632	21·8594
1975	14·2055	14·2095	14·2127	17·4945	17·4905	17·4873
1976	11·4179	11·4153	11·4128	20·2821	20·2847	20·2872
1977	9·1639	9·1693	9·1672	18·0361	18·0308	18·0328
1978	14·5355	14·5277	14·5266	12·6645	12·6722	12·6734
1979	13·2749	13·2709	13·2696	12·1251	12·1291	12·1304
1980	5·7200	5·7200	5·7200	19·6800	19·6800	19·6800
Total	*220·0000*	*215·6699*	*214·4403*	*430·0000*	*430·0000*	*430·0000*

Table 6
Waste Removal Plan (millions of tons, deposit probabilities 10 per cent for each pit)

Year	*Pit 1, Removal probability*			*Pit 2, Removal probability*		
	50%	*10%*	*5·5%*	*50%*	*10%*	*5·5%*
1961	60·0000	60·0000	60·0000	65·0000	65·0000	65·0000
1962	55·4200	56·7280	57·0988	63·8050	66·5148	66·8210
1963	56·5329	56·7070	56·7636	64·2866	64·3435	64·3802
1964	56·2625	56·6038	56·6989	64·0925	64·7960	64·9860
1965	56·3282	56·5692	56·6382	64·1707	64·5233	64·6140
1966	56·3122	56·5407	56·6071	64·1392	64·5472	64·6614
1967	52·8176	56·5202	56·5784	64·1519	64·4822	64·5803
1968	42·2541	51·9719	55·2426	64·1468	64·4707	64·5563

Table 6 (*continued*)

Year	Pit 1, Removal probability			Pit 2, Removal probability		
	50%	10%	5·5%	50%	10%	5·5%
1969	33·8033	41·5775	44·1941	65·7388	66·0283	66·1190
1970	27·0426	33·2620	35·3553	68·3372	68·6379	68·7185
1971	21·6341	26·6096	28·2842	66·0195	67·5977	67·6825
1972	17·3073	21·2877	22·6274	68·2242	67·9723	68·0493
1973	39·9364	41·6777	40·9636	58·1305	60·0946	61·4208
1974	60·2955	59·9743	59·5892	46·5044	48·0757	49·1367
1975	55·8964	50·7606	47·6714	37·2035	38·4605	39·3093
1976	55·8690	40·6085	38·1371	33·8060	43·5872	45·2152
1977	56·4238	60·2295	60·8307	33·4715	31·8418	31·6840
1978	56·2890	55·5537	55·4722	28·0863	30·0782	30·3078
1979	27·1060	23·6804	23·6729	46·6830	49·2779	49·3887
Total	*887·5309*	*906·8621*	*912·4256*	*1066·9976*	*1089·9597*	*1096·6310*

Table 7
Waste Removal Plan (millions of tons, removal probabilities 5·5 per cent for each pit)

Year	Pit 1, Deposit probability			Pit 2, Deposit probability		
	50%	10%	5·5%	50%	10%	5·5%
1961	60·0000	60·0000	60·0000	65·0000	65·0000	65·0000
1962	57·0988	57·0988	57·0988	66·8210	66·8210	66·8210
1963	56·7636	56·7636	56·7636	64·3802	64·3802	64·3802
1964	56·6989	56·6989	56·6989	64·9860	64·9860	64·9860
1965	56·6382	56·6382	56·6382	64·6140	64·6140	64·6140
1966	56·6071	56·6071	56·6071	64·6614	64·6614	64·6614
1967	56·5784	56·5784	56·5784	64·5803	64·5803	64·5803
1968	56·5578	55·2426	53·6902	64·5563	64·5563	64·5563
1969	49·9367	44·1941	42·9522	66·1190	66·1190	66·1190
1970	39·9494	35·3553	34·3618	68·7185	68·7185	68·7185
1971	31·9595	28·2842	27·4894	67·6825	67·6825	67·6825
1972	25·5676	22·6274	21·9915	68·0493	68·0493	68·0493
1973	40·9359	40·9636	41·0181	61·4275	61·4208	61·4259
1974	59·5640	59·5892	59·5856	49·1420	49·1367	49·1327
1975	47·6512	47·6714	47·6685	39·3136	39·3093	39·3062
1976	38·1210	38·1371	38·1348	45·2291	45·2152	45·2207

Table 7 (*continued*)

Year	Pit 1, Deposit probability			Pit 2, Deposit probability		
	50%	10%	5·5%	50%	10%	5·5%
1977	60·8361	60·8307	60·8328	31·6643	31·6840	31·6871
1978	55·4645	55·4722	55·4734	30·2975	30·3078	30·3111
1979	23·6975	23·6729	23·6767	49·3887	49·3887	49·3887
Total	*930·6261*	*930·4256*	*907·2600*	*1096·6312*	*1096·6310*	*1096·6309*

Table 8
Expected Discounted Profits and Shadow Prices

Probability level(%)		Total discounted profits	Shadow prices		
			Deposit	Removal	
Deposit	Removal			Pit 1	Pit 2
50·0	50·0	2130·17	2·1381	0·3654	0·3251
50·0	5·5	2061·58	1·9377	0·3712	0·3330
10·0	50·0	2120·64	2·2182	0·3652	0·3251
10·0	10·0	2068·37	3·0342	0·3700	0·3312
10·0	5·5	2053·02	2·0110	0·3713	0·3330
5·5	50·0	2117·91	2·2222	0·3652	0·3251
5·5	5·5	2050·51	2·0109	0·3713	0·3330

References

CHARNES, A., and COOPER, W. W. (1963), 'Deterministic equivalents for optimizing and satisficing under chance-constraints', *Operations Research*, vol. 11.

VAN DE PANNE, C., and POPP, W. (1963), 'Minimum cost cattle feed under probabilistic protein constraints', *Management Science*, vol. 9, no. 3.

9 R. L. Gue and J. C. Liggett

Mathematical Programming Models for Hospital Menu Planning

R. L. Gue and J. C. Liggett, 'Mathematical programming models for hospital menu planning', *Journal of Industrial Engineering*, vol. 17, 1966, pp. 395–400.

Analytical efforts in the area of hospital menu planning have been rare. Several hospitals in the country have replaced clerical tasks through the use of electronic data processing equipment (see, for example, Rockford, 1963 and Fielding, 1964). These conversions have allowed summary data concerning patient selections to be obtained. More advanced management analysis of the nutrient content of patient meals has been the subject of a few research projects (see, for example, Brisbane, 1964).

New developments in electronic data processing equipment, allowing increased efficiency and flexibility together with advances in mathematical programming have facilitated the development of automated menu planning systems. This article is aimed at outlining the rationale of using mathematical programming in menu planning. Results of the implementation of these concepts in an operating hospital system will also be discussed.

The use of linear programming techniques in planning minimum cost diets has received considerable attention in the past. Much of this work has been summarized by Smith (1963) and will not be covered here. This article will center around planning meals for hospital patients. However, the concepts can easily be applied to other institutional food management systems. Examples of such systems can be found in the military, correctional institutions and schools.

Non-Selective Menu Planning

A daily non-selective menu is obtained by forming a combination of menu items, or dishes, one from each of several menu item classes. A possible set of these menu item classes follows (Balintfy, 1963):

	Menu Item Classes
Breakfast Menu	1. Appetizers
	2. Entrees
	3. Cereals
	4. Breads
	5. Beverages
Dinner Menu	1. Appetizers
	2. Entrees
	3. Starches
	4. Vegetables
	5. Salads
	6. Desserts
	7. Breads
	8. Beverages
Supper Menu	1. Appetizers
	2. Entrees
	3. Starches
	4. Vegetables
	5. Salads
	6. Desserts
	7. Breads
	8. Beverages

The problem of planning non-selective menus is one where over a period of days it is desired to provide a given number of meals (usually three) per day so that:

1. The patient obtains a minimum (or maximum) amount of certain nutrients on each day.
2. There is a reasonable degree of variety among dishes, daily and from day to day.

It must be determined what arrangements of selections should be used for menus during this period in order to minimize, for a prescribed quality, the cost of the menus.

The non-selective menu problem has been stated by Balintfy

(1963, 1964) in two different models. However, one of these seems to have met with greater application and will be the only one presented here. This model is also the one that will be used later in the formulation of the selective menu planning problem.

For a given nutrient, the minimum daily requirement in a menu may be stated mathematically as follows:

$$\sum_{k=1}^{L} \sum_{j=1}^{N_k} d_{ijk}\, v_{jk} \geqslant b_i \qquad i = 1, \ldots, M. \qquad 1$$

In the above, d_{ijk} is the quantity of the ith nutrient in the jth menu item, in the kth menu item class. N_k is the number of menu items in the kth class. L is the total number of menu item classes. The summation is over all items within a class and over all classes. The inequality states that the sum of these quantities must be greater than, or equal to, a certain minimum nutrient level, b_i, $i = 1, \ldots, M$. The variable v_{jk} indicates the number of times a serving of menu item j in class k appears on the daily menu. In this formulation, $v_{jk} = 0, 1$; that is, the item either appears or does not appear on the daily menu plan. Previous research studies have used the minimum dietary allowances as published by the National Research Council for nine nutrients ($M = 9$): calories, protein, calcium, iron, vitamin A, thiamin, riboflavin, niacin and vitamin C (National Research Council, 1958).

The structure of the problem requires that there be one and only one representative from each menu item class. This constraint may be stated symbolically as follows:

$$\sum_{j=1}^{N_k} v_{jk} = 1 \quad k = 1, \ldots, L. \qquad 2$$

It should be noted that the problem is being formulated in terms of planning a single daily menu. However, in actual application it is desired to plan a sequence of daily menus over a period of days (called a cycle). The frequency of serving certain menu items over this period must also be considered. For numerous reasons, many menu items have a minimum time interval, in days, before repetition can occur. This generates a sequential dependency among the daily menus planned over a period of days. This dependency will be discussed later.

The function to be minimized may be defined as

$$z = \sum_{k=1}^{L} \sum_{j=1}^{N_k} c_{jk} v_{jk} \qquad 3$$

where c_{jk} is the cost of the jth item in the kth menu item class. In the analysis of non-selective menus to this time, c_{jk} has represented only raw food cost (Balintfy, 1963; 1964). The primary reason for this is that production cost data are not available in most hospital dietary departments. It should be noted that the models discussed may still be used when these data become available. Since raw food cost makes up the major portion of the cost of the menu, these data provide a good indication of the optimal menu plan. z is the total cost per patient-day for serving the menu.

The problem defined is a zero–one integer programming problem. An integer algorithm has been developed by Balintfy (1963) to yield optimal or near optimal solutions. The algorithm has been programmed for digital computation and approximately thirty seconds is required to obtain a daily menu.

Balas (1965) has developed an algorithm for zero one programming problems which converges to an optimum solution. Applications reported have been with hand computations only. This algorithm has been programmed at the University of Florida for possible use in menu planning.

Healy (1964) has discussed this model in what he calls multiple choice programming. This algorithm, like that of Balintfy, has not been shown to converge to an optimum.

Selective Menu Planning

The problem of planning selective menus is one where over a period of days it is desired to provide a given number of meals (usually three) per day so that:

1. The patient obtains at least a minimum (or, at most, maximum) amount of certain nutrients on each day.
2. The patient is allowed a choice of items from each menu item class.

3. There is a reasonable degree of variety among dishes, daily and from day to day.

The problem is to determine what arrangements of selections should be used in menus during the period in order to minimize, for a prescribed quality, the cost of the menus. This statement is the same as the non-selective menu problem with the important difference that the patient is allowed a selection of items from each class of menu items.

In theory, one may enumerate all possible combinations of patient selections from a given daily selective menu. However, even with certain reductions in the number of menu item classes at the University of Florida, the total number of different possible daily menus makes such an enumeration unfeasible.

Planning selective menus involves choosing menu item groups from a particular class among the menu item classes enumerated earlier. A typical group from the dinner vegetable class might be 'broccoli, carrots'. Broccoli and carrots are examples of menu items. Note that the patient is offered a choice (from a group) in each menu item class. In the following discussion, x_{jk} will represent the jth group of menu items from the kth class of items. x_{jk}^{s} will represent the sth menu item from the jth group in the kth class of menu items. Note $x_{jk} = (x_{jk}{}^{1} \ldots x_{jk}^{s} \ldots x_{jk}^{s})$ where S is the total number of menu items in the group, x_{jk} is the decision variable and will be equal to one if group jk is used in the daily menu and will be equal to zero if it is not used.

Ideally, one would like to know what the patient's choices will be for a given menu plan. However, there is no way of obtaining this information during the menu planning operation. The next best procedure is to make inferences about the patient's choices on the basis of past patient selections. In most hospital situations, this information is readily available. If a given patient is assumed to make a choice of an item from a menu item group with some relative frequency, the repeated observation of this selection would, in the long run, allow one to estimate the probability that the patient will make a particular selection.

Let p_{jk}^{s} be the probability that a patient will choose the sth menu item from the jth group and the kth class of menu items. Assume that when a patient is presented with group $jk(x_{jk} = 1)$, he chooses x^s_{jk} with probability p^s_{jk}. Let a^s_{ijk} be the quantity of the ith nutrient found in the sth menu item in group jk. Thus, when $x_{jk} = 1$, a_{ijk} is a random variable where a^s_{ijk} occurs with probability p^s_{jk}. Let c^s_{jk} be the cost of the sth menu item in group jk. When $x_{jk} = 1$, the cost of x_{jk} is c^s_{jk} with probability p^s_{jk}. Hence, for any x_{jk} there is some probability vector $p_{jk} = (p^1_{jk} \ldots p^s_{jk} \ldots p^S_{jk})$ which gives the probability distribution for a_{ijk} and c_{jk}. p_{jk} is identical for all i. Note that for any menu item class k, S may have a different value. To identify this correctly by S_k would complicate even more what is now a cumbersome notation.

First, the function to be minimized will be considered. The objective function will be similar in form to that of **4** except that the costs, c_{jk}, are now random variables. The approach will be to minimize the expected cost of a daily menu plan. It may be noted that this policy seems to be consistent with a hospital's long run goal to minimize total cost while preserving a specified quality.

One may denote

$$E(z) = \sum_{k=1}^{L} \sum_{j=1}^{N_k} \sum_{c=1}^{S} c^s_{jk}\, p^s_{jk}\, x_{jk}. \qquad \textbf{4}$$

If one lets
$$\bar{c}_{jk} = \sum_{s=1}^{S} c^s_{jk}\, p^s_{jk}$$

one will minimize
$$E(z) = \sum_{k=1}^{L} \sum_{j=1}^{Nk} \bar{c}_{jk}\, x_{jk}. \qquad \textbf{5}$$

$\bar{c}_{jk}$ is the expected cost of menu group jk.

The structure of the selective menu problem requires that one and only one group from a menu class occurs on a daily menu. For example, it is desired that only one dinner vegetable group occur on each daily menu. This requirement is expressed mathematically by the following constraint:

$$\sum_{j=1}^{Nk} x_{jk} = 1 \quad k = 1, \ldots, L. \qquad \textbf{6}$$

If one lets y_i be the quantity of the ith nutrient in a given menu plan, one notes that it is a random variable defined by

$$y_i = \sum_{k=1}^{L} \sum_{j=1}^{N_k} a_{ijk} \, x_{jk} \quad i = 1, \ldots, M. \qquad 7$$

The constraint given by **6** is such that this summation will always contain L non-zero terms. Note that in this expression a_{ijk} is a random variable having values $a_{ijk}^1, \ldots, a_{ijk}^s$. x_{jk} is the decision variable. If one assumes the a_{ijk} are independent for all j and k and that L is sufficiently large, then the Central Limit Theorem (Hogg and Craig, 1965) indicates that the y_i are normally distributed: $n(\mu_i, \sigma_i^2)$. The assumption that the a_{ijk} are independent is not necessarily tenable and will be discussed later. In applications thus far fourteen classes of menu items have been used, $L = 14$.

Let μ_{ijk} and σ_{ijk}^2 be the mean and variance of the random variable a_{ijk}. Then,

$$\mu_i = \sum_{k=1}^{L} \sum_{j=1}^{N_k} \mu_{ijk} \, x_{jk},$$

$$\sigma_i^2 = \sum_{k-1}^{L} \sum_{j-1}^{N_k} \sigma_{ijk}^2 \, x_{jk}^2. \qquad 8$$

If one lets b_i be the minimum daily requirement of a patient for the ith nutrient, one may express the probability that the patient's requirement is not met by

$$\int_{-\infty}^{bi} n(\mu_i, \sigma_i^2) dy_i. \qquad 9$$

The implicit requirement in planning a patient's menu is that all his minimum daily nutrient requirements be met. Due to the random nature of the selective menu planning problem, meeting this requirement with cetainty is impractical. Alternatively, a minimum acceptable probability that the patient's requirement for the ith nutrient will not be met has been specified. Mathematically, the following constraint has been formulated:

$$\int_{-\infty}^{bi} n(\mu_i, \sigma_i^2) dy_i \leqslant a_i \quad i = 1, \ldots, M \qquad 10$$

where a_i is the maximum acceptable probability of not meeting the patient's requirement for nutrient i.

An additional constraint required is

$$x_{jk} = 0, 1 \quad \text{all } _{j,k} \qquad \mathbf{11}$$

where $x_{jk} = 0$ indicates that menu group jk was not used in the daily menu plan and $x_{jk} = 1$ indicates the group was used. The problem is to minimize **5** subject to the constraints given by **6**, **10** and **11**. The problem defined is a stochastic zero-one programming problem (Hadley, 1964).

Table 1
Typical Estimates of Selection Probabilities and Menu Item Costs for the Breakfast Appetizer Class

Group	*Estimated probabilities* ($p_{jk}{}^8$)	*Item costs* ($C_{jk}{}^8$)
Grapefruit Sections	0·55	0·384
Prune Juice	0·30	0·200
Strawberries	0·33	0·330
Orange Juice	0·48	0·510
Plums	0·23	0·283
Grapefruit Juice	0·58	0·143

In planning daily menus for a period of days (a menu cycle) it is usually desirable to place a lower bound on the time between the reappearance of a particular menu item, x^s_{jk}, the sth item in the jth group in the kth class. For example, it may be desirable to serve the dinner entree of fried chicken no more often than every five days. It is not possible to formulate this requirement as a mathematical constraint in planning a daily menu. The computer program developed for solution of the problem monitors this minimum interval and will not allow a group to enter the solution if it contains a menu item that does not satisfy the interval restriction (Gue and Liggett, in press).

It should be pointed out that planning a sequence of daily menus individually over a period of days, say, twenty-one, is not equivalent to planning the entire menu cycle at the same time. In fact, planning the daily menus individually may be considered suboptimization. Also, most hospitals have selective therapeutic diets in addition to a regular diet. All these menus must be planned simultaneously for true optimality. Problem formulation and the development of algorithms for these more complex problems is still underway at the University of Florida (Gue, 1966).

Algorithm for Solution

Two algorithms for solution of the problem defined have been considered. An extension of Balas' algorithm has been developed and is being programmed for the I.B.M. 709. An extension of Balintfy's algorithm has been developed and programmed for the I.B.M. 709. All applications up to this time have used this extension. Details of the algorithm will not be covered here. Those wishing a more detailed description should see Gue and Liggett (in press).

Results

The parameters found in **4** have been estimated from data taken at the University of Florida. The costs, c_{jk}^{s}, were estimated from purchasing records. The probabilities, p_{jk}^{s}, were estimated from approximately 4000 patient selections. Table 1 gives typical values of p_{jk}^{s} and c_{jk}^{s} for two groups of menu items. It is not possible to write **4** explicitly since the problem contained well over 200 decision variables.

The b_is found in **10** were based on minimum daily nutrient requirements for a twenty-five year-old male as set by the National Research Council (1958).

These minimum allowances are shown in Table 2. At present, nine nutrients have been considered, but others may be easily added to the analysis. One should note that eggs, beverages, condiments and bread items have been omitted from the menu planning operation. Normally, the same selection of these items is offered to the patient every day. Hence, they are not subject to menu planning. The nutrient content of these items, shown in

Table 2, is subtracted from the minimum allowances to obtain the b_is shown.

Table 2
Minimum Dietary Allowances Used in Selective Menu Planning

Nutrient	*National Research Council allowances*	*Adjustment*	*Adjusted allowances* (b_i)
Calories	2600	691·5	1908·5
Protein	70 (gm.)	36·1	33·9
Calcium	800 (mg.)	860·8	—*
Iron	10 (mg.)	2·6	7·4
Vitamin A	5 (1000 I.U.)	1·1	3·9
Thiamin	1·0 (mg.)	0·7	0·3
Riboflavin	1·6 (mg.)	1·4	0·2
Niacin	17 (mg.)	1·9	15·1
Vitamin C	70 (mg.)	6·6	63·4

*The calcium requirement is met by eggs, beverages and condiments.

The nutrient values of individual menu items were based on food composition data published by the United States Department of Agriculture. The mean nutrient content of a group was estimated by its definition using estimated probabilities p^s_{ijk} :

$$\hat{\mu}_{ijk} = \sum_{s=1}^{S} a^s_{ijk} p^s_{ijk} .$$

Typical nutrient values for two groups are shown in Table 3.

Recall that the variance given by **8** assumes independence of the random variables a_{ijk}. It is important to note that in some cases patient selections are not independent. For example, a patient's selection of a dinner vegetable may depend on the meat dish he chooses. Therefore, a Central Limit Theorem for dependent random variables was used and the summation given by **7** was still assumed to be normally distributed. In estimating the variance, σ_i^2, appropriate covariance terms were included.

The time required for planning a daily selective menu, using the extension of Balintfy's algorithm, is approximately forty seconds. Up to this time, a fourteen-day cycle of regular, non-therapeutic diets has been planned for the University of Florida Teaching Hospital. These diets account for a major portion of the

Table 3

Typical Estimates of Nutrient Content for the Breakfast Appetizer Class

Group	($a_{1_{jk}}{}^{s}$) *Calories*	($a_{2_{jk}}{}^{s}$) *Protein*	($a_{3_{jk}}{}^{s}$) *Calcium*	($a_{4_{jk}}{}^{s}$) *Iron*	($a_{5_{jk}}{}^{s}$) *Vitamin A*	($a_{6_{jk}}{}^{s}$) *Thiamin*	($a_{7_{jk}}{}^{s}$) *Riboflavin*	($a_{8_{jk}}{}^{s}$) *Niacin*	($a_{9_{jk}}{}^{s}$) *Vitamin C*
Grapefruit Sections	30·00	0·60	13·00	0·30	0·01	0·03	0·02	0·20	30·00
Prune Juice	77·00	0·40	15·00	0·30	0·05	0·05	0·02	0·20	9·00
Strawberries	92·00	0·40	13·00	0·60	0·03	0·02	0·06	0·50	55·00
Orange Juice	46·00	0·60	10·00	0·20	0·20	0·08	0·03	0·40	44·00
Plums	63·00	0·40	9·00	0·90	1·23	0·02	0·00	0·40	2·00
Grapefruit Juice	33·00	0·50	8·00	0·40	0·01	0·03	0·02	0·20	31·00

total volume of meals served at the Hospital. A typical output of the selective menu planning program is shown in Table 4. Also shown is the expected raw food cost to the hospital for serving each of the menu groups, as well as the total expected raw food cost for the entire day. Note that eggs, beverages and bread items have been omitted from the menu plan. In order to get a true indication of the total expected raw food cost, the calculated cost of the menu plan must be added to the cost of beverages, bread items and condiments.

Selective menus planned with the models discussed in this article were served to patients on a pilot basis, during a two-week period in August 1965. Exploratory interviews with patients were conducted and indicated no adverse reactions to these menus.

Conclusion

Estimated savings of menus planned with mathematical programming as opposed to menus planned with traditional methods have ranged from six to 25 cents per patient-day (Balintfy, 1963; Gue and Liggett, in press; Liggett, 1965). As far as the authors can determine, all applications thus far have been in hospital systems. The estimated savings in the selective menu system at the University of Florida Hospital (approximately six cents per patient-day) are considerably less than those estimated for non-selective menus at Tulane (Balintfy, 1963). It is felt that one of the reasons for this difference could be the relative efficiency of the traditional method for menu planning currently being used at Florida. Secondly, the uncertainties caused by random variation in the selective menu planning model prevents the precision in the selective menu planning that may be obtained in planning non-selective menus.

A break-even analysis made in the University of Florida (Liggett, 1965) and the Tulane studies (Balintfy, 1963) indicate that the use of mathematical programming and a digital computer to plan hospital menus may allow major savings. Full implementation of these methods is underway but yet to be achieved. Hopefully, the future will bring adequate evidence of the practical value of the use of mathematical programming in menu planning.

Table 4
Selective Menu Planned by Computer

	Expected cost in cents
Breakfast	
Prunes, Orange Juice	3·089
Sausage	3·828
Wheat Chex, Oatmeal, Grits	0·692
Cherry Preserves	0·799
Dinner	
Beef Noodle Soup	0·638
Fried Chicken, Baked Pork Chop, Mashed Potatoes, Sweet Potatoes/Apples	24·069
Broccoli, Carrots	3·129
Tossed Salad/Blue Cheese Dressing, Congealed Pear Salad	4·007
Boston Cream Pie, Royal Anne Cherries	3·370
Supper	
Mushroom Soup	0·572
Macaroni and Cheese, Seafood Casserole	8·717
Green Beans, Blackeye Peas	3·274
Chef's Salad/French Dressing, Pineapple Cheese Salad	3·157
Coconut Cake, Fruit Cocktail	2·670
Total expected cost	62·011*

* This cost does not include the cost of eggs, bread, beverages and condiments which could add as much as 45 cents.

References

BALAS, E. (1965), 'An additive algorithm for solving linear programs with zero-one variables', *Operations Research*, vol. 13, no. 4.

BALINTFY, J. L. (1963), 'Mathematical programming of menu planning in hospitals', Tulane University School of Business, New Orleans, (monograph.)

BALINTFY, J. L. (1964), 'Menu planning by computer', *The Communications of the ACM*, vol. 7.

BRISBANE, H. M. (1964), 'Computing menu nutrients by data processing', *J. Am. Dietetic Assoc.*, vol. 44, no. 6.

FELLERS, J. D., and GUE, R. L. (1965), 'Computer planning and control of dietary functions', Paper delivered at the Annual Meeting, American Hospital Association, San Francisco.

FIELDING, V. V. (1964), 'Computer tell what to put on the menu', *Modern Hospital*, vol. 100, no. 4.

GUE, R. L. (1966), 'Linear programming models for planning selective hospital menus', *Proceedings*, AIIE, San Fransisco.

GUE, R. L., and LIGGETT, J. C. (in press), 'Selective menu planning by computer' J. Hillis Miller Health Center, University of Florida, Monograph.

HADLEY G. (1964), *Non-Linear and Dynamic Programming*, Addison Wesley

HEALY, W. C., Jr (1964), 'Multiple choice programming', *Operations Research*, vol. 12, no. 1.

HOGG, R. V., and CRAIG, A. T. (1965), *Introduction to Mathematical Statistics*, Macmillan.

LIGGETT, J. C. (1965), 'Mathematical programming and hospital menu planning: Applications and Limitations', Master's Thesis, University of Florida.

NATIONAL RESEARCH COUNCIL, (1958), *Recommended Dietary Allowances*, N.R.C., Publication no. 589, Washington, D.C.

SMITH, V. E. (1963), *Electronic Computation of Human Diets*, Michigan State University Business Studies, East Lansing.

ROCKFORD (1963), 'Data processing speeds tallying of selective menus at Rockford (Illinois) Hospital', *Hospitals*, vol. 37, no. 12.

10 P. Mellor

A Review of Job Shop Scheduling

P. Mellor, 'A review of job shop scheduling', *Operational Research Quarterly*, vol. 17, 1966, pp. 161–71.

Introduction

The purpose of this paper is to summarize and discuss recent contributions to the formulation and solution of scheduling problems centring on the task of sequencing a set of jobs on a group of machines in what is generally described as a job shop situation.

The point is well made by Brown (1960) that it is not helpful to identify a problem as a 'job shop scheduling' problem because 'ingenious' solutions have been found by exploiting special characteristics of a particular problem not found in the large class of problems of which it is a member. Hence a somewhat extensive knowledge of which characteristics have been successfully exploited, and in what manner, is necessary if the literature is to be put to work.

However, it is not intended that this paper will become a mere catalogue of honourable mentions, but, rather, the broad theme which connect many papers will be stressed, the relationship of 'academic' formulations to real problems discussed, and the awareness of many problem solvers of the importance of the context of their problems emphasized.

Reviews of the literature and comprehensive bibliographies on job shop scheduling problems and closely allied fields have been published by Sisson (1959; 1961) and Thompson (1960) and a brief review in Muth and Thompson (1963), in works which are readily available, and it is not my intention to go over this ground again. But the customary list of references to papers discussed has been augmented by all the relevant articles of which I am aware and which are not covered by the earlier reviews.

Definition of the Problem

Job shop scheduling is a field in which the idiosyncratic definition of terms is commonplace and it is customary for every paper to begin with such a definition which differs, ever so slightly, from every other definition. Scheduling itself is a word so rich in different, and confusing, connotations that an attempt will be made to avoid any particular use of it in the rest of this paper. It seems to be becoming fairly generally accepted that the central scheduling problem can be most usefully described as *sequencing* (it has also been called 'dispatching' – for example, Conway, Johnson and Maxwell, 1960), the entities which pass through the shop are called *jobs* in this paper (the other common name is 'commodities') and the work done on them at a *machine* ('facility') is here called an *operation* ('task'). Where it is applicable, the required technological ordering of the operations on each job is called a *routing*. These definitions will be used consistently in the paper and in the main follow Gere (1962) though quotations from reviewed papers have not been edited into line. Thus the sequencing function can be defined (for the moment, non-operationally) as the ordering of the operations on jobs at the machines, subject to routine constraints, so that the best value is obtained for some measure of effectiveness appropriate to the system. The ordering which is obtained is called a *programme* (geometrically it would be a Gantt chart) and this can be produced by the sequencing operation as a projected time-table or may only become apparent retrospectively if a system of priority numbers is used dynamically, or if jobs are introduced dynamically into the shop.

Models of job shops which have been proposed invariably have had highly restrictive shop disciplines, and it has been the concern of many researchers to investigate the effectiveness of sequencing procedures when certain restrictions have been relaxed (see, for example, Muth (1963) on the consequences of uncertainty about the durations of operations).

The following list of typical simplifying assumptions is taken from Gere:

(1) No machine may process more than one operation at a time.

(2) Each operation, once started, must be performed to completion (no pre-emptive priorities).
(3) Each job, once started, must be performed to completion (no order cancellations).
(4) Each job is an entity; that is, even though the job represents a lot of individual parts, no lot may be processed by more than one machine at a time. This condition rules out assembly operations.
(5) A known, finite time is required to perform each operation and each operation must be completed before any operation which it must precede can begin (no 'lap-phasing'). The given operation time includes set-up time.
(6) The time intervals for processing are independent of the order in which the operations are performed. (In particular, set-up times are sequence-independent and transportation time between machines is negligible.)
(7) In-process inventory is allowable.
(8) Machines never break down and manpower of uniform ability is always available.
(9) Deadlines (due dates), if they exist, are fixed.
(10) The job routing is given and no alternative routings are permitted.
(11) There is only one of each type of machine (no machine groups).
(12) All jobs are known and are ready to start processing before the period under consideration begins. (This is the 'static' scheduling problem.)

This list is based on Sisson (1961) where a summary table of the applicability of many of these assumptions to various classes of models is contrasted with their applicability to a 'typical actual situation'.

As would be expected, the more abstract models for which more rigorous, analytical solutions have been obtained, for which algorithms to obtain optimal solutions exist, exhibit more of the more severe restrictions, but many of these are relaxed in real problems which have been studied and for which simulation models have been built in attempts to obtain 'acceptable' solutions by observing the effects of priority rules or by using heuristic methods.

Most sequencing systems can be put into one of two classes according to whether their objective function is aimed at completing all the jobs in a static set as soon as possible – irrespective of the time at which any individual job is completed – or optimizing the value of some system parameter as the equivalent criterion in cases where jobs are continually introduced into the shop; or whether their aim is to optimize the value of some function of the time at which individual jobs are completed. These two classes can be loosely described as the 'minimum make-span' problem and the 'due date' problem.

Typical criteria cited by Gere for the first class are:

(1) Finish the last job as soon as possible; that is, minimize the interval of time from the start of processing until all jobs are completed.
(2) Finish each job as soon as possible. (Minimize the sum of completion times.)
(3) Minimize the in-process inventory costs.
(4) Maximize machine utilization –

and for the second class:

(1) Minimize the number of late jobs.
(2) Minimize the total tardiness.
(3) Minimize the costs due to not meeting due dates exactly. This includes the first two as particular forms of the cost function. Costs due to earliness might be included.

Of course, mixtures of these criteria have been used, particularly by researchers investigating by simulation the effectiveness of priority rules for resolving conflicts at a particular machine for both classes of problem. For example, Conway, Johnson and Maxwell (1960) use the following measures of performance:

(1) The distribution of times to complete a job – time from introduction to the shop to the completion of processing on the last operation.
(2) The distribution of lateness of jobs – the length of time between the actual completion of a job and the desired completion.
(3) The amount of work-in-process inventory in the shop.
(4) The utilization of shop facilities – the complement of idle time.

In this work, these measures of performance are considered

separately and not combined by the assignment of cost coefficients. Beenhakker (1963a) considers the difficulties of measuring the effectiveness of a sequence when 'there is no common measure of value for the various desirable properties' and uses Utility Theory to construct an appropriate pay-off function. His list of system goals which can be attained by good sequencing ranges very widely:

(1) Minimum idle facility investment.
(2) Minimum in-process inventory.
(3) Minimum facility set-up costs.
(4) Day-to-day stability of work force.
(5) Adherence to promised shipping date.
(6) Maximum output (production rate).
(7) Minimum materials – handling cost.
(8) Adherence to arbitrary job priorities, such as arise in dealing with preferred customers, emergency repair parts, etc.
(9) Technological feasibility.
(10) Sensitivity to possible production changes.
(11) General flexibility.
(12) Non-dependence on unreliable processes.
(13) Reserve capacity for rush orders.
(14) Optimal in-plant transportation schedule.
(15) Minimum shipping costs.
(16) Minimum total expected costs, primarily in theoretical investigations.
(17) Maximum weighted facility utilization.
(18) Maximum utilization of manpower.
(19) Optimal assignment of various labour grades.
(20) Minimum raw material inventory.
(21) Minimum finished product inventory.
(22) Minimum investment in inventories.
(23) Minimum obsolescence and deterioration of products.
(24) Shortest make-span for certain products.
(25) Minimum overall fabrication span.
(26) Minimum risk of excessive losses.
(27) Anticipated changes in price.

Although, in real life, most of these factors are of great concern to management, needless to say no one, to my knowledge, has even attempted to devise a sequencing system which combines them, operationally, to establish a pay-off function. To do so would involve the solution of practically every scheduling problem which has ever been studied, if not, indeed, the solution of every operational research problem. Rather, as can easily be inferred from the list of simplifying assumptions given earlier, it is usual to so define one's model that most of these real life complexities are excluded and so that job sequencing does not become involved with investment problems, batch size and run-length problems, scheduling for assembly (CPM, etc.) or assembly-line balancing, and rarely are sequence dependent set-ups (asymmetrical travelling salesman problems) considered in this context. Such a procedure can readily be defended as the path of progress for otherwise intractable problems, but there is a danger of the tractable problems becoming somewhat unreal.

The most important point, however, is that good sequencing has many side-effects and that, in real problems, the context of the sequencing function has to be considered explicitly if, to quote Sisson, one is to 'optimize[s] a lesser criterion chosen in some reasonable way'. This is discussed in the next section.

Context of the Problem

The very difficult combinatorial problems which have either to be avoided or solved in job shop scheduling have obvious attractions for many operational research people working at the theoretical level. It is not so obvious that industrial management is particularly concerned with obtaining optimum solutions to these problems, or, indeed, that people in industry who are involved in scheduling are particularly conscious of the difficult sequencing problems which they are supposedly solving.

Pounds (1963) reports his disconcerting experience of being unable to find anyone in industry who was responsible for the detailed sequencing of jobs who recognized that he had a scheduling problem. He infers from his experience that:

> The job-shop scheduling problem is not recognized by most factory schedulers because *for them*, in most cases, no scheduling problem exists. That is there is no scheduling problem for them because the

organization which surrounds the schedulers reacts to protect them from strongly interdependent sequencing problems.

In effect, industrial schedulers are being asked to get a pint out of a quart pot and are experiencing no difficulty in doing so. The scheduler's protection of course, comes from the extravagant provision of shop capacity or poor commercial performance. Pounds concludes:

> If those parts of the organization which now control the constraints in such a way as to avoid difficult scheduling problems could learn to react, not to present limits of scheduling ability, but to the new limits implied by the more powerful techniques – then and only then will progress have been made.... The benefits to the firm of better scheduling algorithms will not result from changes in the scheduling activity alone. Unless policy is changed in the areas of capital budgeting, labour cost control, employment policy, sales policy, and policies on customer service, the new scheduling techniques ... could conceivably do more harm than good....

Thus it is suggested that the operational criterion of the sequencing process might not be an appropriate measure of the effectiveness of the changes made to the scheduling procedure; and, further, that the efficiency of sequencing should be explicitly considered when tackling problems in the areas affected by good or bad sequencing.

It is possible to think of the three areas of capacity planning, plant loading (batch sizes, quotation of delivery dates, etc.) and job sequencing as presenting separate problems, but it is also possible that the most interesting problems arise from the interactions between them. There is a brief discussion of some of the problems in this area in Mellor (1965).

In these circumstances it is not surprising that few cases of implementation – or even proposals for implementation in any particular case – have been reported on. Gere (1962) reports some inconclusive trials using real data; Guest and Tocher (1963) describe a scheduling project concerned with sequencing (though not strictly for a job shop); and there is some discussion of sequencing problems in an industrial context in Acton, Robinson and Tobin (1963). Most of the accounts that one hears of computer-based scheduling systems seem to concentrate on the information-handling and data-processing functions and accept the existing sequencing methods as satisfactory. Yet it is largely because of the impact of computers on management thinking and the stress laid by operational research on the dangers of suboptimization that some encouragement is now being given to the better solution of these sequencing problems.

Some Solutions to the Problem

It is a commonplace that, despite the power of modern computers, the one method of solution that is not feasible in any real-sized problem is to lay out all possible sequences and then pick the best. For J jobs and M machines, in the general case there will be $(J!)^M$ such sequences.

Similar combinatorial problems exist in most fields of decision making and Brown (1960), commenting on the difficulties, draws an analogy with the difficulties of computing optimum strategies for playing real games. He takes the first three plays of a simplified version of a very simple game (Gops) and quickly generates $10^{20,000}$, possible strategies. His apt comment is: 'It is almost unbelievable that such an innocent-appearing little game could generate this kind of combinatorial madness.'

Despite all this, Giffler, Thompson and Van Ness (1963) report on computational experience with the complete enumeration of all active feasible schedules (as previously defined by Giffler and Thompson, 1960) when operation times are all set at unity and Monte Carlo versions for both the non-numerical (unity) and numerical cases.

Brown stresses the importance of experience, imagination and intuition used by people who are actually faced with these combinatorial problems and concludes that computers can help in these situations but hardly replace humans. Vazsonyi (1965) has described how human–computer interfaces might be constructed to operate in this kind of situation.

Other researchers have taken the view that computers can be programmed to operate in the same way as humans, but can solve problems more consistently and expeditiously. Dutton (1964) has specifically set out to simulate the behaviour of a human scheduler who can analyse and classify incoming orders into a few types. Much of the work currently proceeding on heuristic programming is highly relevant (see, for example, Minsky (1961) for an early but comprehensive description).

A certain amount of work has been done in grafting heuristic devices on to sequencing systems based on priority numbers. An attempt to make such a system 'learn from experience' is reported by Fischer and Thompson (1963). They conclude:

(1) An unbiased random combination of scheduling rules is better than any of them taken separately.
(2) Learning is possible.

but question whether the slight advantage obtained by learning justifies the extra computation.

Gere has studied the performances of a number of combinations of priority rules and rules of thumb for the 'due date' problem with both static and dynamic order books. Gere selects his priority rules on the basis of work done by Rowe (1960), Baker and Dzielinski (1960) and Conway and Maxwell (1963); they are:

(1) Job slack.
(2) Job slack per operation.
(3) Job slack ratio (job slack hours divided by hours remaining until the due date).
(4) 'Modified' job slack ratio, which takes machine loading into account by adding to each operation time the expected delay time associated with it.
(5) Length of next operation. This is the 'SIO' rule (shortest imminent operation).
(6) Length of next operation together with job slack ratio under certain circumstances.
(7) First come, first served.
(8) Random.

The rules of thumb are such as seem intuitively reasonable as likely to improve a sequence based on the blind adherence to a priority rule. The devices tested are:

(1) *Alternate operation.* Schedule the operation according to the rule. Now check to see if this makes another job 'critical' (that is, see if the slack of any other job has just become negative, or if positive, has reached a certain critical level). If so, revoke the last operation,

and schedule the next operation on the critical job. Check again for lateness. If scheduling the second operation does not cause any job to be critical, whereas the first one did, then schedule the second one; otherwise schedule the first one (that is, the one which was dictated by the rule).

(2) *Look ahead.* When an operation is scheduled we may ask: Is there a critical (i.e. late or nearly late) job due to reach this machine at some future hour, yet before the scheduled operations is completed? If so, schedule that job. Check to see the effect of this on other jobs. Either let the schedule remain, or replace the operation with the previously scheduled operation, depending on the resulting job lateness effected. . . .

(3) *Insert.* Once a 'look ahead' job has been scheduled there is a period of idle time on the machine, starting at the 'present' time. If there is a job in the queue whose next operation can be completed by the time the look ahead job is due to arrive at this machine, then the operation should be scheduled. . . .

After testing these rules and heuristic devices, Gere somewhat cautiously concludes:

(1) When the problem under study is the realistic one with the goal of meeting delivery commitments, the selection of a priority rule for discriminating between jobs competing for time on the machines is not as important as the selection of a set of heuristics which will bolster the rule.

(2) Since there is little difference in effectiveness of the priority rules after they are combined with two or more heuristics, a simple rule should be used.

Two simple rules are the job slack and shortest imminent operation rules. At the other extreme rather extensive calculations are entailed in the modified job slack ratio rule which takes machine loads into account. Except for an experimental research program, complex priority rules are inadvisable.

(3) The heuristics which anticipate the future progress of a schedule, the alternate operation and look ahead heuristics (together with the insert routine), improve schedules significantly in both a statistical and practical sense, and the improvement in performance is ample reward for the incremental computing cost.

The principal alternative approach to using local rules based on priority numbers is to find an efficient method of generating a global programme for all the operations on all the jobs in the job set at all the machines; this approach is particularly appropriate to the minimum 'make-span' problem with a static order book.

The most attractive method conceptually is to use an integer linear programming formulation. Such an approach has been described by Bowman (1959), Wagner (1959) and Manne (1960) among others. Unfortunately, computational experience has so far proved disappointing. Story and Wagner (1963) conclude a report on experiments with a three-machine model with the words:

> We have not yet found an integer programming method that can be relied upon to solve most machine sequencing problems rapidly. We believe that future study must concentrate on deriving methods which more fully take into account the special structure of machine sequencing problems. . . .

Perhaps the recent progress in integer programming reported by Beale (1965, p. 316 et seq., in this volume) holds out more hope for the future.

An approach which has proved computationally more efficient is to set out to generate all feasible schedules in such a way that an optimum or near-optimum solution is found after generating only a small subset of the possible sequences. Recent progress with this method stems from the branch-and-bound algorithm for the travelling salesman problem developed by Little *et al.* (1963) but the method is only applicable to the three-machine case (see the recent paper by Lomnicki, 1965) or where the order of processing jobs is constrained to be the same on all machines – an equivalent case; a four-machine example of this has been investigated by Ignall and Schrage (1964).

An equally attractive alternative approach is to generate only such a subset of sequences as can be shown to include the optimum sequence – or a sequence equivalent to the optimum from the 'minimum make-span' point of view. The seminal paper here is that by Giffler and Thompson (1960) in which they develop the concept of 'active, feasible schedules' and an algorithm for generating all active schedules. Unfortunately, for many real problems, it is impracticable to generate even this subset of possible sequences and then one is obliged to generate only a sample of the subset. It has been shown by Beenhakker (1963b) that by random sampling with replacement there is a higher probability of selecting sequences with a smaller 'make-span' than those with a longer 'make-span', and he goes on to develop a decision rule for establishing the optimal number of schedules to generate, taking into account the 'total return of a schedule' and the cost of producing schedules by this method.

A geometric model, presented by Hardgrave and Nemhauser (1963), is similar to the active schedule approach. Here again it seems advantageous to construct bounding paths before carrying out calculations to determine an optimal path.

There is one further general topic which is currently receiving attention and that is the effect of uncertainty on the goodness of schedules. Muth (1963) has studied the effects of errors in job times and suggests other sources of error:

(1) Other job requirements (e.g. equipment and materials needed), particularly human skills required.
(2) Engineering changes.
(3) What the job actually turns out to be (e.g. in repair work).
(4) What the objectives of a scheduling system are.
(5) The extent to which variables may actually be chosen by the decision maker.

Despite his conclusion that '... the schedule span is not very sensitive to moderately large errors in estimated job times', it seems intuitively reasonable that the better the information that is used in constructing a sequence the better that sequence will be. This is supported by most of the work on sequencing using local priority rules, where better results have been obtained if the priorities are re-evaluated every time a decision has to be taken. The work of Guest and Tocher (1963) is essentially an attempt to cope with high variability by generating a fresh schedule every time a decision is required and only implementing those parts of it which require immediate action.

Of course, this approach requires elaborate mechanisms for keeping the scheduler's knowledge of the state of the shop up to date and adequate computing capacity for on-line real-time operation. Particular, and apparently largely unresolved, difficulties would seem to arise if it is desired to combine the skill of human schedulers with this method of computer scheduling – i.e. submit computer-generated sequences to human review – and still operate the total system on-line.

Conclusion

There have been many other 'ingenious' solutions proposed in recent papers which have not been discussed in this review, but I have attempted to include all of them of which I am aware in the bibliography. In addition, of course, many of the works listed have excellent bibliographies.

My main concern in this review has been to illustrate how recent developments appear to offer practicable methods of solution for some real problems, but at the same time to stress the dangers of presenting solutions to limited sequencing problems without recognizing the contextual problems which might overwhelm proposals for implementation. It seems that whilst real progress has undoubtedly been made with idealized, highly abstract models, this progress has not been reflected in the solution of real problems – or perhaps such progress has simply not been reported on.

References

ACTON, R. A., ROBINSON, D., and TOBIN, N. R. (1963), 'Developing an automatic scheduling system for a new integrated steelworks', *Proc. Third Int. Conf. Opl Res., Oslo*, Dunod, Paris (1964).

BAKER, C. T. and DZIELINSKI, B. P. (1960), 'Simulation of a simplified job shop', *Management Science*, vol. 6, p. 311.

BEALE, E. M. L. (1965), 'Survey of integer programming', *Operational Research Q.*, vol. 16, p. 219.

BEENHAKKER, H. L. (1963a), 'Development of alternative criteria for optimality in the machine sequencing problem', Ph.D. Thesis, Purdue University, Indiana.

BEENHAKKER, H. L. (1963b), 'Mathematical analysis of facility-commodity scheduling problems', *Int. J. Prod. Res.*, vol. 2, p. 313.

BOWMAN, E. H. (1959), 'The schedule-sequencing problem', *Operations Research*, vol. 7, p. 621.

BROWN, G. W. (1960), 'Computation in decision making', in R. F. Machol (ed.), *Information and Decision Processes*, McGraw-Hill.

CONWAY, R. W., JOHNSON, B. M. and MAXWELL, W. L. (1960), 'An experimental investigation of priority dispatching', *J. Indust. Eng.*, vol. 11, p. 221.

CONWAY, R. W., and MAXWELL, W. L. (1963), 'Network scheduling by the shortest operation discipline', *Industrial Scheduling*, Prentice-Hall, ch. 17.

DUTTON, J. M. (1964), 'Production scheduling – a behavioural model', *Int. J. Prod. Res.*, vol. 3, p. 3.

FISCHER, H., and THOMPSON, G. L. (1963), 'Probabilistic learning combinations of local job-shop scheduling rules', *Industrial Scheduling*, Prentice-Hall, ch. 15.

GERE, W. S., Jr (1962), 'A heuristic approach to job shop scheduling', Ph.D. Thesis, Carnegie Institute of Technology.

GIFFLER, B., and THOMPSON, G. L. (1960), 'Algorithms for solving production-scheduling problems', *Operations Research*, vol. 8, p. 487.

GIFFLER, B., THOMPSON, G. L., and VAN NESS, V. (1963), 'Numerical experience with the linear and Monte Carlo algorithms for solving production scheduling problems', *Industrial Scheduling*, Prentice-Hall, ch. 3.

GUEST, G., and TOCHER, K. D. (1963), 'The control of steel flow', *Proc. Third Int. Conf. Opl Res., Oslo*, Dunod, Paris (1964).

HARDGRAVE, W. W., and NEMHAUSER, G. L. (1963), 'A geometric model and a graphical algorithm for a sequencing problem', *Operations Research*, vol. 11, p. 889.

IGNALL, E., and SCHRAGE, L. (1964), 'Application of the branch and bound technique to some flow shop scheduling problems', *Operations Research*, vol. 13, p. 400.

LITTLE, J. D. C., MURTY, K. D., SWEENEY, D. W. and KAREL, C. (1963), 'An algorithm for the travelling salesman problem', *Operations Research*, vol. 11, p. 972.

LOMNICKI, Z. A. (1965), 'A branch-and-bound algorithm for the exact solution of the three-machine scheduling problem', *Operational Research Q.*, vol. 16, p. 89.

MANNE, A. S. (1960), 'On the job shop scheduling problem', *Operations Research*, vol. 8, p.219.

MELLOR, P. (1965), 'Applications of linear programming in steelworks planning', Operational Research Society, unpublished paper.

MINSKY, M. (1961), 'Steps towards artificial intelligence', *Proc. I.R.E.*, vol. 49, p. 8.

MUTH, J. F. (1963), 'The effect of uncertainty in job times on optimal schedules', *Industrial Scheduling*, Prentice-Hall, ch. 18.

MUTH, J. F., and THOMPSON, G. L. (1963), Introduction and Section Prefaces to *Industrial Scheduling*, Prentice-Hall.

POUNDS, W. F. (1963), 'The scheduling environment', *Industrial Scheduling*, Prentice-Hall, ch. 1.

ROWE, A. G. (1960), 'Towards a theory of scheduling', *J. Indust. Eng.*, vol. 11, p. 125.

SISSON, R. L. (1959), 'Methods of sequencing in job shops – a review', *Operations Research*, vol. 7, p. 10.

SISSON, R. L. (1961), 'Sequencing theory', in R. L. Ackoff (ed.), *Progress in Operations Research*, Wiley.

STORY, A. E., and WAGNER, H. M. (1963), 'Computational experience with integer programming for job-shop scheduling' *Industrial Scheduling*, Prentice-Hall, ch. 4.

THOMPSON, G. L. (1963), 'Recent developments in the job-shop scheduling problem', *Naval Res. Logistics Q.*, vol. 7, p. 585.

VAZSONYI, A. (1965), 'Automated information systems in planning control and command', *Management Science*, vol. 11, p. B2.

WAGNER, H. M. (1959), 'An integer programming model for machine scheduling', *Naval Res. Logistics Q.*, vol. 6, p. 131.

Additional References

DUDEK, R. A., and TEUTON, O. F. Jr (1964), 'Development of *M*-stage decision rule for scheduling *n* jobs through *M* machines', *Operations Research*, vol. 12, p. 471.

EASTMAN, W. L., EVEN, S., and ISAACS, I. M. (1964), 'Bounds for the optimal scheduling of *n* jobs on *m* processors', *Management Science*, vol. 11, p. 268.

GIFFLER, B. (1963), 'Schedule algebras and their use in formulating general systems simulations', *Industrial Scheduling*, Prentice-Hall, ch. 4.

LAWLER, E. L. (1964), 'On scheduling problems with deferral costs', *Management Science*, vol. 11, p. 280.

LEVY, F. K., THOMPSON, G. L., and WIEST, J. D. (1963), 'Mathematical basis of the critical path method', *Industrial Scheduling*, Prentice-Hall, ch. 22.

MAXWELL, W. L. (1964), 'The scheduling of single machine systems: a review', *Int. J. Prod. Res.*, vol. 3, p. 177.

NANOT, Y. R. (1963), 'An experimental investigation and comparative evaluation of priority disciplines in job shop-like queuing networks', Ph. D. Dissertation, University of California.

PALMER, D. S. (1965), 'Sequencing jobs through a multi-stage process in the minimum total time – a quick method of obtaining a near optimum', *Operational Research Q.*, vol. 16, p. 101.

WAGNER, H. M., GIGLIO, R. J., and GLASER, R. G. (1964), 'Preventive maintenance scheduling by mathematical programming', *Management Science*, vol. 10, p. 316.

11 S. E. Elmaghraby and R. T. Cole

On the Control of Production in Small Job Shops

S. E. Elmaghraby and R. T. Cole, 'On the control of production in small job shops', *Journal of Industrial Engineering*, vol. 14, 1963, pp. 186–96.

Our experience with several small shops producing a variety of products indicates that, in general, the problems of production control in such shops can be listed as follows:[1]

1. There is an absence of a reliable estimate of current or future shop loads and the necessary shop capacity to cope with them. It is not uncommon for a shop to subcontract a sizeable portion of its future orders with the expectation of full utilization of capacity to find, when the time comes for the manufacture of the unsubcontracted orders, that the shop is greatly under-utilized.

2. Employment is relatively unstable. Clearly, this is partially caused by the first condition mentioned above. However, even after giving due allowance to the changing nature of the products, and the fact that the shop is a component of a larger system – which is often the case – it is obvious that the shops are subject to severe and sometimes irrational manpower fluctuations.

3. It is difficult for the shops to respond as fast as desired to customer demands concerning either design or quantity changes. This is found to be mainly due to the lack of knowledge by the shop management of the current status of the various orders in the shop.

4. The manufacturing interval (defined as the total time the job is in the shop) is relatively long, which causes dispatchers to resort to 'rush' activities a great deal of the time to expedite the movement of urgently needed jobs. Needless to say, these activities cause other jobs to be delayed, and the vicious cycle continues.

In one instance, for example, we found that in spite of the fact that, on the average, a lot took approximately twenty hours of

1. Almost identical problems exist in large shops.

work, the average manufacturing interval was quoted as eight weeks. The difference between the few hours required and the weeks that must pass before jobs are finished represents idle time during which the jobs are waiting as in-process inventory.

5. The shops usually experience *both* overtime and idle time expenses. The reason for the simultaneous occurrence of these two conflicting phenomena can be found in the fact that *imbalances* in the loads on the various operations are coupled with the *inability* of management *to know* of these imbalances fast enough to rearrange the manpower distribution. This would necessarily result in one operation being overloaded with work (and hence resorts to overtime) while another operation is 'starving' for work (and hence experiences idle time expenses).

6. The presence of a relatively large number of clerks and dispatchers employed in each shop and, at the same time, lack of co-ordination between the *loading* function and the *dispatching* function is another problem.

7. A relatively large number of jobs exist in the shop at any time, far in excess of normal load-balancing requirements. These jobs contribute heavily to the in-waiting investment, to management's lack of knowledge of the whereabouts of jobs, to the load of dispatching and physical inventory taking, and last but not least, to the size of books and files kept by the first line supervision.[2]

Objectives of the System

The diagnosis of the areas of deficiency enumerated leads, in a natural way, to the realization of the need for control systems that provide:

1. A scheduling procedure that effectively regulates the amount of work released to the shop in any given interval of time, *taking into account the actual shop capacity and performance.*

2. An estimate of required manpower for current and future activities.

2. We are not certain of the causal relationship here – whether the lack of control results in the large number of jobs in the shop, or whether the existence of many jobs helps to confuse management and foredoom any attempts to control!

3. A fast and accurate status reporting system that keeps management in constant touch with the behavior of the shop under the changing load conditions.

The Interaction Between the System Design and Computer Hardware

Before proceeding with the details of the construction of the control system, we first discuss a basic concept concerning the configuration of the 'hardware' to be used in implementing the system.

It is easy to see that it is not enough to specify *what* should be done without specifying *how it can* be done. This brings to the forefront the question of the interaction between the design of the control system and the computer-hardware system that is the carrier through which the procedures are to be implemented. We believe that this question has been either glossed over or totally neglected heretofore.

For example, in the application reported upon in this article, some thirty operations existed in the shop. The total number of 'jobs'[3] in the shop at any time is approximately 500 (corresponding to about 220 engineering codes). Clearly, if we are to be able to maintain an 'on-line' partial status reporting system, which was one of the requirements of the control system, then the quantity of units in the lot, the route as well as the identification of each job must all be kept on the drum memory of the computer (for the sake of immediate access). This would enable us to update the memory by the transmissions received from the input devices and be able to answer any query concerning the progress of the job or the quantity of units surviving.

The restriction to a small computer, that is, a rather small memory capacity, forced the following to be considered:

1. The need for numbering the jobs in the shop different from the code (that is, design) number, since the latter was up to seven alpha-numeric characters.

3. A 'job' is created whenever an engineering design (or 'code') is started in production. Since the same code may be started in more than one lot (or due to split lots), the number of jobs is usually larger than the number of codes.

2. The re-grouping of the thirty odd operations into 'stations' not exceeding fifteen in number.
3. The limitation of the lot size to a maximum of 1999 units per lot.

The second consideration profoundly affected the ability to *schedule* the shop taking the capacity of the various operations into consideration. In fact, the identities of certain operations were necessarily lost in the process of grouping, and *no capacity* could be assigned to the resulting heterogeneous station.

We reiterate, at the risk of being redundant, that the design of the control procedure is directly affected by the vehicle of its implementation and vice versa. The same interaction between the equipment and the control system design was evident in every phase of the control procedures. It was evident in the design of the input devices, in the reports issued by the system, and in the very construction of the scheduling procedure.

The Scheduling Procedure

We have termed the specification of the codes to be *started in a fixed interval of time*, the determination of the size of the production lots as well as the designation of their priorities as 'The Shop Schedule'. Our use of the word is parallel to Magee's definition of the functions of scheduling, and is close to its common usage in the shop concerned (Magee, 1958).

The problem of scheduling-sequencing job shops in general has been treated by Wagner (1959), Bowman (1959) and Manne (1960). The generality of the treatment of these authors – which, in essence, amounts to the construction of an optimal Gantt chart – is paid for rather heavily in the complexity of the mathematical formulation of the problem (as integer linear program) and the fact that it is not feasible to obtain schedules for any but the most trivial problems in terms of the number of products and the size of the shop.

The determination of the *economic manufacturing quantity* for each code is easily seen to be intimately related to the utilization of available capacity in each individual station, since to manufacture in excess of immediate demand (in order to achieve the desired economies of manufacture) is contingent upon the availability of shop (that is, individual station) capacity. Clearly management's desire is to achieve *both* objectives: *manufacture in economic quantities and fully utilize available capacities*. Oftentimes, these two objectives cannot be realized simultaneously and a compromise is necessitated. Attempts to achieve an *optimal* compromise have been made by Manne (1958), Rogers (1958) and Dzielinski, Baker and Manne (1961). However, the computations involved in all three proposals preclude their utilization in small operating systems such as we are concerned with here.

'Unit' shops

Perhaps the most elementary but non-trivial problem in this context is the problem of scheduling 'unit shops', that is, shops that are scheduled as one unit and not subdivided into smaller 'stations' or 'operations'. It is also the case most widely discussed in the literature (Dzielinski, Baker and Manne, 1961; Manne, 1958; Rogers, 1958). In many instances, such as in-process industries, the production facilities lend themselves easily to the unit shop concept. In other instances, such as outlined in the following section, certain characteristics of the shop force such an interpretation.

Let

r_{ij} denote the customer demand of code i in period j, measured in units of the product, $i = 1, \ldots, N$; $j = 1, \ldots J$.

h_{ij} the total production time for r_{ij}, that is, $h_{ij} = r_{ij}t_i + s_i$ where t_i is the manufacturing time (excluding set-up time) of a unit, and s_i is the set-up time.

$\hat{h}_{ij} = h_{ij} - s_i$

l hourly labor and machine cost.

c_i cost of storage per unit of code i per scheduling period.

C capacity of the shop.

X_{ij} a variable that takes on the value 1 or 0 according to whether r_{ij} is included in current schedule or not.

N the number of codes on demand.

It is obvious that if

$$\sum_i h_{i1} \geqslant C$$

that is, if the total requirements in the *current* (that is, first) period are equal to or greater than the available capacity, no pooling of future deliveries with current production is possible. (In fact, resort to overtime may be necessitated). Therefore, our discussion is restricted to the case where the relationship in **1** does not hold.

Conceptually, all the codes on order are in competition for the available net capacity $(C - \sum h_{i1})$. If the demand in the jth period, r_{ij}, is pooled with current production, we *save one set-up* and incur the *cost of storing* r_{ij} for $(j-1)$ periods. Hence, the net economic advantage is

$$a_{ij} = s_i \times l - (j-1)c_i \times r_{ij}. \qquad \mathbf{2}$$

Thus, the problem of producing economic manufacturing quantities subject to the limitation on available capacity can be mathematically formulated as

$$\text{Maximize} \quad \sum_i \sum_j a_{ij} X_{ij} \qquad \begin{array}{l} i = 1, \ldots, N \\ j = 2, \ldots, J \end{array} \qquad \mathbf{3}$$

$$\text{Subject to} \quad \sum_i \sum_j \hat{h}_{ij} X_{ij} \leqslant C - \sum_i h_{i1} \qquad \mathbf{4}$$

$$\text{And} \quad X_{ij} \leqslant X_{i(j-1)} \qquad \mathbf{5}$$

The last restriction insures that pooling will proceed in ascending time periods without 'jumps'.

This is an *integer* linear program (LP) with $N(J-2)+1$ restrictions and $N(J-1)$ variables. The magnitude of the computational aspects of this formulation can be gleaned from the following consideration. A shop with 300 codes and demand that runs six months into the future would yield a system of 1201 equations in 1500 unknowns (excluding 'slack' variables). The number of unknowns and restricting equations can be drastically reduced by taking into consideration some obvious facts. For example, we are certainly not interested in any $a_{ij} < 0$ since this implies an economic *disadvantage*. Therefore, a preliminary calculation, which can be easily performed on a desk calculator, would reveal the expensive codes whose 'pooling' horizon is less than J. Again, prior knowledge of the instability of design, expected customer changes, the experimental nature of the code or any other reason that may interfere with the desirability of producing more than the immediate demand limits the pooling horizon of some codes to the first period, and further reduces the number of unknowns and the number of restrictions of type **5**.

The solution of the integer LP in **3**, **4** and **5** yields an optimal schedule. The optimality of the schedule is predicated on the following two assertions:

1. It is more economical to produce whole requirements rather than partials. Clearly, if the shop produces $(r_{i1} + r_{i2} + \ldots + ar_{ij})$ in period 1, where $0 < a < 1$, then in period j the remainder, $(1 - a)r_{ij}$, will have to be produced. Obviously, the cost of storing $a.r_{ij}$ in $(j - 1)$ periods could be saved by producing the total quantity, r_{ij}, at period j, *without incurring any extra set-up costs.*

2. It is uneconomical to produce while storing (Wagner and Whitin, 1958). That is, if r_{ij} is pooled for production with r_{i1}, then *all* intervening demand, $r_{i2}, r_{i3}, \ldots, r_{i(j-1)}$, must also be in the same pool. This follows from the fact that if r_{ik}, say, is not in the pool, $k < j$, then it will have to be produced in period k. But then, we could pool – at no extra set-up cost – all the requirements between periods k and j, $r_{i(k-1)}, \ldots, r_{ij}$, in the same lot and would thus save the storage cost

$$\sum_{m=1}^{j-k} r_{i(k+m)} \times (k - 1) \times c_i .$$

Approximations to the LP formulation – marginal analysis

Consider a job shop which is divided into homogeneous stations for reasons of reporting status and other accounting considerations. Suppose that most of the work done, whether manual or mechanical, requires approximately the same grade of workers. Then, clearly, the shop management would achieve greater flexibility by 'manning' the various stations according to the load requirements at any particular time. In such a situation the workers are usually trained in more than one operation and they are shifted constantly from one station to the other. Therefore the shop can be treated as a 'unit' shop.

Let

$$F_{ij} = I_i \times a_{ij}/(h_{ij} - s_i). \qquad 6$$

F_{ij} is the 'Pooling Preference Factor' (PPF); it measures the marginal utility (per hour) of pooling the *j*th demand with cur rent production. I_i is an index; it is equal to 1 or 0 depending on whether the future deliveries of the code are to be pooled or not, thus permitting prior selection of the 'poolable' codes. The *j*th delivery of code *v* is pooled in preference to the *k*th delivery of code *w* if

$$F_{vj} > F_{wk}. \qquad 7$$

The relationship in 7 gives the key to the procedure of pooling. For each code we calculate the PPF of the second delivery with the first. (Note that if, for some code *i*, $r_{i1} = 0$ its PPF would be < 0. Also the second delivery may be required in some period $j > 2$.) The PPFs are then ranked in order of decreasing magnitude, and the highest-ranking code is pooled first. Immediately thereafter, its PPF for the *next* delivery is calculated, and the cycle of ranking and pooling repeated. The process of pooling ends when:

1. The excess capacity is reduced to near zero.
2. There are no more deliveries to pool.
3. When all F_{ij} are $\leqslant 0$.

The degree of approximation (using the marginal analysis approach) depends to a large extent on the ratio of Max $(h_{ij} - s_i)$ to the capacity *C*; the smaller the ratio the better the approximation.

If at the end of every pool we plot the economic advantage versus the hours loaded, the result would be a polygon, such as (*A*) in Figure 1. Note that the polygon is necessarily concave. In the case of large values of $(h_{ij} - s_i)$ relative to the total capacity *C*, the polygon is as in (*A*): few deliveries are pooled, a large capacity is unutilized, and the economic advantage is low. On the other hand, small values of $(h_{ij} - s_i)$ would yield polygon (*B*) terminating at point E_2, which represents higher economic gains and better capacity utilization.

In our application of this procedure, the ratio of Max $(h_{ij} - s_i)$ to the shop capacity was of the order of 0·003 to 0·01. This insured us of a good approximation to the optimal solution.

The dispatching priority

The mathematical models of the two preceding sections result in the identification of the codes to be released to the shop, the period of their manufacture and the quantity to be started in each period. Since 'a period' may be of considerable length – in our application it was one month – it is desirable to assign a 'starting priority' to the various codes. This priority indicates the *order* of release to the shop. Clearly, in a company where service is of paramount importance, the priority rule should be predicated on meeting the customer due dates.

Moreoever, it is evident that the determination of the 'release' priority of a code is a *static* function which is unsatisfactory from the point of view of day-to-day operation. Since priority is a *relative* measure (although it may be expressed in a cardinal rather than ordinal scale), it should be *dynamic* in character to reflect, as much as possible, the continuously changing relative position of the code.

The priority rule adopted for this application is a function of the *slack time* and is *updated at each status reporting period*, which is much shorter than the scheduling period. Thus, the priority P_i for code i is given by

$$P_i = \left\{ D_i - D_0 - \left\lceil \frac{H_{ij}(k) + A(k)}{8} \right\rceil \right\} \qquad 8$$

where D_i is the due date of the earliest delivery in a pool; D_0 is the current date; $H_{ij}(k)$ is the operation time in the *remaining* operations through which the code must pass; and $A\ (k)$ is the allowance for interference and unavoidable delays, and is a function of the *remaining* stations in the route. In this application, $A\ (k) = 12 + 8k$ hours. The square brackets [ ] stand for 'the smallest integer larger than or equal to the quantity inside the brackets.' Note that D_i and D_0 refer to the 'working' days rather than regular calendar days.

Note that if two codes have identical due dates but one is to pass through more operations than the other, its priority would be higher. On the other hand, two jobs of identical route and standard operating time would have different priorities depending on their due dates.

The *dynamic* aspect of this rule is reflected in the fact that D_0, $H_{ij}(k)$, and $A(k)$ change with the passage of time and the progress of the job in the shop. It is also apparent that should a job remain in some operation without being worked on, D_0 will progressively get larger and P_i will consequently decrease thus causing the job to have higher priority.

The slack time rule adopted in this study is one of several priority rules that have been proposed in the literature. The formulation of **8** has special merit since P_i actually measures the available slack time (to due date) in multiples of eight-hour day periods. A positive P_i indicates the presence of slack time; and a negative P_i indicates 'back-schedule' conditions. The existence of a physical meaning to the number printed on the various reports is a valuable asset in rendering the priority rule understandable (and hence 'practical') to the shop personnel. A more detailed evaluation of the behavior of the rule in practice is given in a later section.

At the moment, we wish to state one final remark concerning dynamic priority rules in general, and in particular rules of the 'slack time' variety.

In all such rules some allowance must be made for the *average* queueing time at each operation. Such delays are inherent in job shop operations because of the indeterminate nature of the time standards on the one hand, and the desire to avoid machine and operator idle times on the other. However, the very fact that a job has a high (or low) priority means that its waiting time is *very different from the average*; thus, the priority is a function of the anticipated delay time which is, itself, a function of the priority! Thus, rigorously speaking, **8** should appear as follows

$$P_i = D_i - D_0 - \left[\frac{H_{ij}(k) + f(P_i, k)}{8}\right] \qquad \mathbf{9}$$

where $f(P_i, k)$ is the delay allowance which is a function of P_i as well as k, the number of future operations.

The analysis of this problem is beyond the scope of this presentation. However, it is interesting to note that only recently has attention been given to the fact of the interdependence between the priority and some of the factors that determine it (Sandman, 1961).

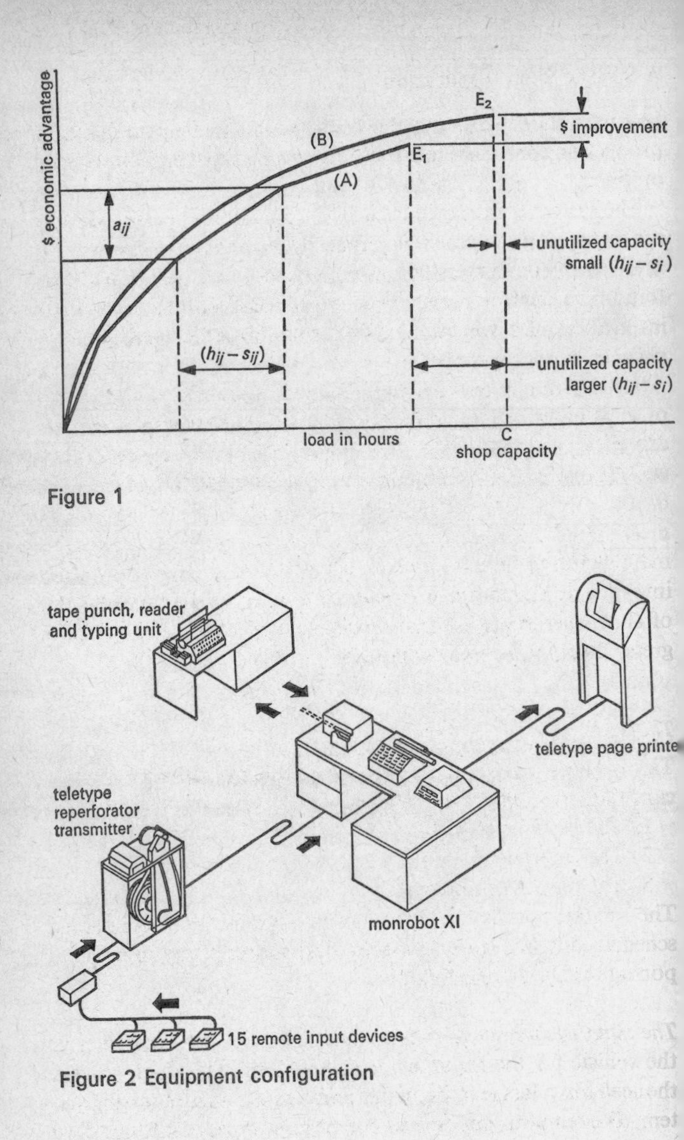

Figure 1

Figure 2 Equipment configuration

A Specific Shop Application

In this section we detail the construction and operation of a production control system that implements the theory previously outlined.

The system is managed exclusively by the shop personnel and the computing equipment is physically located in the production area. The shop is classified as a job shop, and demands a fast responsive control system because of the short manufacturing interval coupled with a high frequency of design and quantity changes by the customer.

The exterior system design is directed toward the optimization of local criteria. The boundaries of the system were defined to exclude subsystems such as raw material ordering and shop accounting, and other information or control systems related to the present system through product intercourse. The principal measures used to judge system effectiveness were reductions in the manufacturing interval and raw material as well as in-waiting investment. Significant improvements in control and balancing of shop operations, estimation of machine loads and job progress control were also demanded.

The equipment

The equipment used for the implementation of the control system can be divided into three categories:

1. The input equipment.
2. The computer.
3. The output equipment.

The data-transmission-computation system is represented schematically in Figure 2; a brief description of the three components of the system follows:

The input equipment. A remote data collection system serves as the vehicle for the input of job movement. This tool satisfies the need for a fast reacting status and performance reporting system. The remote input devices accept three types of information:

1. Fixed numeric information, namely, the lot number entered by a card permanently punched with numeric information.

2. Variable numeric information, namely, the quantity processed which is entered through four dials.

3. Three buttons provide for entry of information fixed by electrical circuitry within the device. The entry in this system is the identification of the work station, each button representing a different station.

4. In addition to the acceptance of these three types of information, provision is made for the transmission of a single character which is, in this system, interpreted by the computer program station transmission. Any other information, requiring only this on-off content, could be inserted provided that the computer is programmed to interpret it correctly. For example, the fact of machine failure could be so recorded immediately at the computer.

It should be noted that the meaning of any of the entries is dependent primarily upon how the central processing unit (that is, the computer) handles the data. The opportunity for system flexibility is, therefore, not seriously impaired by the simplicity of the data collection equipment design.

Certain features were introduced in the design of the input devices in an attempt to minimize operator error. For example a strict operator sequence is required for proper device operation. Analysis of the data available over a year's period indicates that the operator error rate is approximately 5 per cent. This compares favorably with experimental results reported by Deininger under less severe operating conditions (Deininger, 1960).

The input devices are connected to a central scanning device that is continuously monitoring the remote input devices to determine if any 'request for service' condition exists. When such a condition obtains, the input device is connected to a paper tape punch.

As the jobs progress through the shop, operators transmit status data from the stations. Upon the completion of an operation (or several operations grouped in one reporting station), the operator enters the job number, the quantity processed and the station number through the remote input device. These transmissions are recorded on the paper tape punch and stored as a buffer paper tape prior to machine processing. The paper tape punching operation functions to:

1. Allow alternate computer usage, such as printing, without interruption of normal shop traffic.
2. Permit audition of transmissions.
3. Reconstruct the computer memory and tape records in the event of equipment failure.

An important element of the input – as well as the output – equipment is an off-line combined tape punch, reader and typing unit. This piece of equipment prepares all the system paper tapes other than the tape generated by the remote data collection system, and provides an off-line printer for two major reports.

The computer. The central processing unit for the control system is a Monrobot XI computer.[4] It is a digital, binary computer with a nonvolatile drum memory of 1024 words of 32 binary digits each. This computer is extremely versatile in its input as well as output functions. Input to the computer can be accomplished through a typewriter, punched paper-tape readers, punched card readers, Teletype paper-punch machines as well as through sixteen-key numeric keyboards. The specific system reported upon here uses a typewriter, two paper tape readers and a Teletype reader. Possible output devices are typewriters, paper tape punches, paper card punches and Teletype machines. In this system, output is made through the typewriter, a Teletype page printer and the computer paper tape punch.[5]

The output equipment. As is mentioned previously, the output equipment consists of a teletype page printer, the typewriter and the off-line tape punch; as well as punched paper tape. The system configuration is such that the Teletype page printer does the on-line printing (that is, accepts, through electrical connexions, data directly from the computer), while the off-line tape reader accepts computer-punched paper tape as its input.

The flow of information and system reports

Upon receipt of a customer order, analysis is made to determine the need for engineering analysis. Unengineered products are routed to the engineering department for the determination of

4. Monroe Calculating Company.

5. The typewriter, one paper tape reader and one paper tape punch are standard equipment with the computer.

manufacturing instructions. Following the addition of the standard time values by the Wage Incentive organization, this document is sent to the production control system for processing. A dual path, leading directly to the system, also exists for previously engineered products.

Standard time estimating function. The information needed by the control system, such as the manufacturing route, hours and cost data, is often seriously delayed by the engineering analysis cycle. Early estimates of these factors would allow better estimates of delivery commitments to customers as a function of uncommitted capacity and more accurate short range machine loading forecasts. A least-squares multiple regression estimator was developed based on four predictor variables. This statistical model provides time estimates from which other necessary values may be derived. The validity of the model was tested with the standard statistical tests on samples of shop data. A similar application with eight predictor variables was reported by Canning (1959). A reduction of engineering interval of approximately 80 per cent was achieved through this estimation.

The 'customer load' report. Either the manufacturing information or regression factors are transmitted to the operator of the off-line tape reader who prepares an edge-punched card containing all the manufacturing information intended to remain fixed for the life of the product. Monthly, these fixed records are merged with customer demand schedules on punched paper tape. This tape is processed in the computer to result in the first major system report – the customer load report. This report serves to provide management with a total status of customer orders, by quantities and total hours for six months.

The 'shop schedule'. Following the issuance of the customer load report, the computer merges the information in the customer load report with the 'in-waiting' quantity data and available shop capacity. The product is a monthly Shop Schedule. This report is shown in Table 1. Column 1 indicates the letter '*U*' for unengineered codes; such codes are not usually pooled because of their design instability. Column 2 gives the code, or identification, number of the product. The letter '*P*' appears in column 3 if the code is pooled. Column 4 shows the net pooled quantity. Column 5 provides the quantity to be started which is the required quantity plus shrinkage allowance. Column 6 shows the number of units being processed in the shop. Column 7

Table 1
Shop Schedule

							Month			
	Code number		*Re-quired*	*Start*	*In wait-ing*	*Priority*	*2*	*3*	*4*	*5*
	046217		12	20	16	11				
	10C731		32	42		6				
	10C732		20	28	20	6				
	10C733	*P*	40	60	10	10				
	10D588		225	324	300	8				
	10D672		41	56	40	6				
	130330		10	16	8					
U	13E148		840	1182		– 6	274	440	120	240
U	16A247		21	36		– 3				
	16A248		43	68		1				
	16A249		40	64	60	– 5				
	16A250		49	76		– 5				

provides the starting priorities and the remaining four columns show the residual number of units required for four months. The priority and economic lot-size calculations are both explained in a previous section. The report is an objective guide for management to balance efficient shop loading with satisfactory delivery dates. A notable measure of flexibility in control is attained through the ability to:

1. Alter the pooling interval according to changes in shop manufacturing interval.

2. Provide shop management with the prerogative to alter available shop capacity according to desired loading conditions and interval.

The release of jobs to the shop is accomplished from the schedule-generated priorities. Continually during the month, the operator prepares paper tapes containing the fixed information from the edge-punched cards with the due date and schedule-specified lot quantity. These paper tapes are called 'new jobs' tapes and serve, when merged with previous new job tapes, as the master file record in the computer system. This master file tape contains the manufacturing information, quantity and date requirements for all jobs in the shop. (A memory address on the computer drum serves as the lot identification number for the shop personnel as well as computer inquiry.) The extensive use of this external memory device (namely, punched paper tape), permits a high degree of computer utilization and, in effect, an expansion of its nominal capacity.

Daily the current drum information is used to update the master file tape. The master file tape is thus modified to reflect the shop activity, and the information necessary for the system reports is also calculated.

The reports based upon this daily updating procedure are:

1. The Shop Station Report.
2. The Job Status Report.
3. The Station Priority Listing.
4. The Shipping Report.

The shop station report. This report is shown in Table 2. The 'stations' numbered from 1 to 15 are shown in the first column. Opposite each station is given, consecutively: The 'output' in number of units and standard hours expended (columns 2 and 3), and the 'shrinkage' due to defective production in number of units and standard hours (columns 4 and 5). Column 6 gives the standard hours' content of the jobs waiting in the queue *at* the station – hence the heading 'Station Queue Hours'. Column 7 gives the standard hours' content of all the jobs in process in the shop that will be processed by the station, whether they are physically queueing at the station or at some other station. The significance of these two columns should be apparent since the difference between the two entries for any station represents the 'work to come' to the station, and is a good indication of the required manpower. For example, the 'Queue' and the 'Load' at station 1 are equal, indicating that *all* the jobs that are to be worked on in station 1 are physically present at the station.

If we compare this situation with station 13, we notice that the work in the queue is a small fraction of the load on this station. Therefore, the shop supervisor should either expedite jobs passing through station 13, or otherwise reduce the manpower in that station until the queue hours increase appreciably. This report is printed daily and is intended to provide a close control over output and efficiency as well as present a picture of the short range work commitment. Normal operating capacity figures have been established for the loading stations and the load and queue figures refer to these capacities, which are manually inserted in the last column.

The job status report. The Job Status Report, printed semi-weekly on the Teletype page printer, provides full status information about every job in the shop and is shown in Table 3.

Each lot in the shop has a job number (a three-digit number) and a code number (a six-digit number). When a lot is released to the shop for manufacturing, all the entries in the body of the report are the standard times of processing at the various stations. For example, job 229 (code 402179) is worked on in stations 1, 2, 3, 5, 6, 7, 10, 11 and 15; the processing time in station 1 is 0·06 hour, in station 2 is 0·60 hour, etc.

The quantity of units started is normally in excess of customer requirements by the allowance for 'dropout' or defective units. Thus, for job 229, the 'quantity due' is 100 units, but the 'quantity started' is 144 units, which is 44 per cent in excess of end requirements. The third quantity appearing in the same column represents the quantity surviving, based on the latest status transmission. For job 229, which is a new job just released to the shop, the 'quantity remaining' is equal to the 'quantity started', namely, 144 units. But such is not the case for job 214 (code 351035) where the quantity surviving is two units *less* than the quantity started.

As the job progresses through the shop, the 'hours required' are replaced by the 'quantity completed' which is indicated between brackets. This artifice provides a complete up-to-date historical picture of the progress of the job in the shop, helping to pinpoint stations of excessive 'dropout'. For example, consider job 190 (code 352276). The 16 units released to station 1

Table 2
Shop Station Report
Period

	Output		*Shrinkage*		*Station queue hours*	*Station load hours*	*Station capacity hours*
Station	*Units*	*Hours*	*Units*	*Hours*			
1	7330	3	10		31	31	
2	219	5			27	91	
3	106	5			32	288	
4	2736	5			26	200	
5	5117	14			29	153	
6	3407	9			24	212	
7	4648	17	140	1	75	510	
8	225	26	3	1	33	100	
9	643	22	71	1	23	183	
10	4282	55	225	1	204	1608	
11	3770	155	392	35	429	3203	
12	46	17			1156	2384	
13	196	23			2	984	
14	2601	15	13		94	881	
15	2064		467		272	1958	
T	*39390*	*370*	*1321*	*39*	*2458*	*12785*	

Note: Hours are rounded off to nearest integer; the absence of hours indicates a small fraction that was rounded off to zero.

have been reported intact through station 7. In station 8, however, only 15 units were reported, indicating a 'dropout' of *one unit*. Station 9 reported 11 units – implying that 4 units were defective. The quantity started was 16 units – the quantity remaining is only 11 units, and the customer requires 7 units.

The remaining three entries in the Job Status Report occupying the right-most three columns are: the 'job priority' which is updated according to the movement of the job; the 'next station' which should coincide with the indication from the body of the report; and finally the 'total hours remaining' which is simply the sum of the standard hours in the body of the report.

Two special features of the report are important to emphasize. First, whenever a lot is split, the 'splinter' is given a new job number which appears with the 'parent' every time the latter is printed. For example, consider job 223 (code 320439). The asterisk after the job number indicates that it is a 'split job,' and its 'splinter' is job 348 which appears immediately below the job. This method facilitates cross-referencing various jobs which originally started as one job. It is also necessary in order to calculate the 'quantity required' and 'quantity surviving' for the original (parent) job. These quantities appear in column 8.

Second, whenever the 'quantity remaining' falls *below* the 'quantity due', an indication is made in the report to highlight this fact and point out the need for 'restarts'. This is evidenced in job 249 (code 320451): the 'quantity remaining' is only 14 units, which is 6 units *less* than the 'quantity due' which is 20 units. The semicolon appearing next to the 'quantity started' high lights this fact and draws immediate attention to the need for corrective measures.

The station priority report. The Station Priority Report listing, Table 4, is printed on the typewriter simultaneously with the printing of the Job Status Report. It extracts certain information from the Job Status Report for use by the shop operators. For each load station it provides the quantity and priority of all the jobs queued there. For example, station 7 has jobs 269 (4 units), 369 (72 units) and 437 (18 units) whose priorities are 39, 6 and − 3 respectively. It is intended as a guide to sequencing jobs in the shop according to their priority urgency.

The shipping report. The final report, the Shipping Report, provides a daily list of the code number and quantity of boards shipped from final inspection. Its use is found in internal auditing and as an updating record to the production service organization. In addition, a punched paper tape containing the identical

Table 3
Job Status Report
Project Design

Job no. *Code no.* *Split from no.*	*Quantity completed at station in parentheses. Hours required to complete station no parentheses*					*Quantity of units: started, due, remaining*		*Slack time*	*Next station*	*Total hours remaining*
	1 2 3	4 5 6	7 8 9	10 11 12	13 14 15	*Job*	*Total*			
190										
352276	(0016)	—	(0016)	001·23	—	0016				
	(0016)	—	(0015)	000·10	—	0007				
	(0016)	(0016)	(0011)	000·58	000·25	0011		—07	10	002·16
214										
351035	(0060)	—	(0058)	004·54	000·30	0060				
	(0060)	—	009·48	001·64	—	0050				
	(0058)	(0058)	—	—	001·30	0058		—15	08	017·26

Job no. Code no. Split from no.	*Quantity completed at station in parentheses. Hours required to complete station no parentheses*					*Quantity of units: started, due, remaining*		*Slack time*	*Next station*	*Total hours remaining*
	1 2 3	4 5 6	7 8 9	10 11 12	13 14 15	*Job*	*Total*			
223*										
320439	—	—	(0010)	(0010)	(0010)	0010				
	(0010)	—	(0010)	(0010)	(0010)	0007	0015			
348,	(0010)	(0010)	—	(0010)	000·81	0010	0020	−12	15	000·81
229										
402179	000·06	—	000·28	004·27	—	0144				
	000·60	000·05	—	002·68	—	0100				
	001·74	000·17	—	—	000·96	0144		−06	01	010·81
249										
320451	(0015)	—	(0014)	(0014)	005·22	0025;				
	(0015)	—	(0014)	004·77	001·36	0020				
	(0015)	(0014)	—	001·82	000·97	0014		−11	11	014·14

* Split Job.

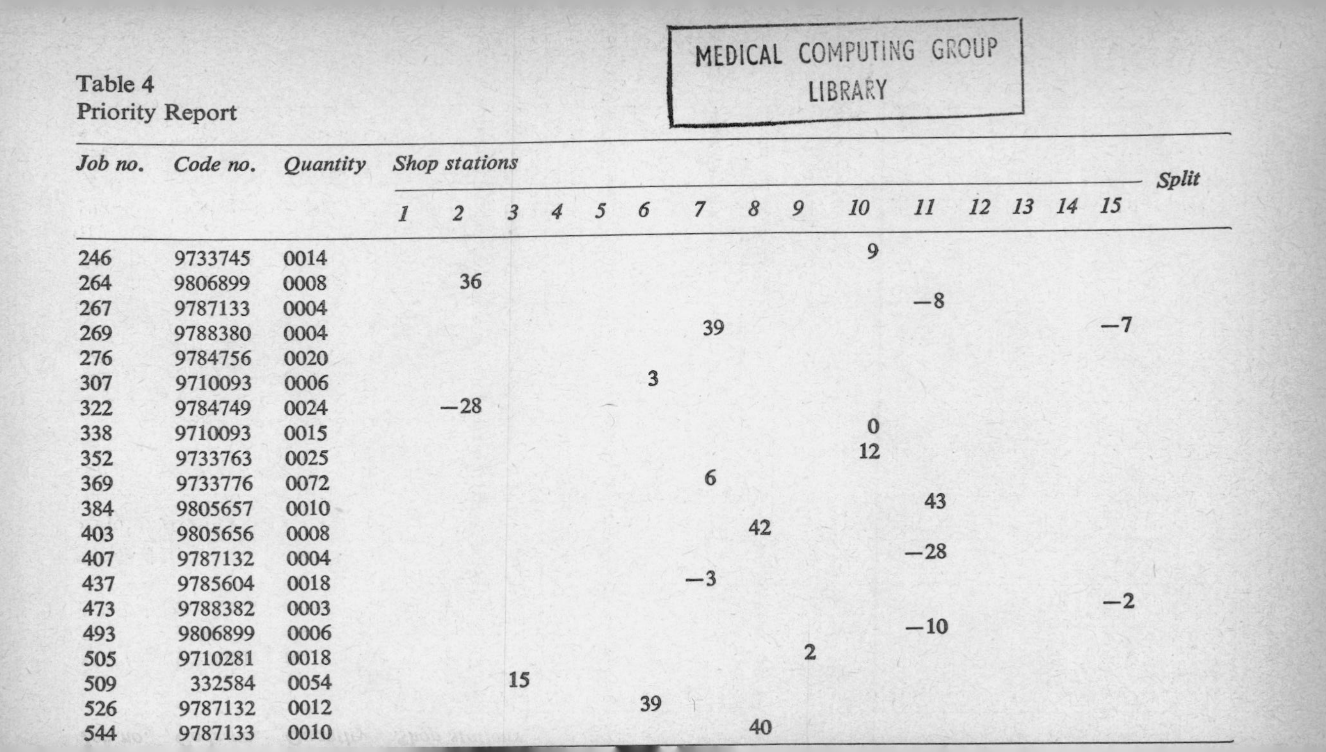

Table 4
Priority Report

Job no.	*Code no.*	*Quantity*	*Shop stations*															*Split*
			1	*2*	*3*	*4*	*5*	*6*	*7*	*8*	*9*	*10*	*11*	*12*	*13*	*14*	*15*	
246	9733745	0014										9						
264	9806899	0008		36														
267	9787133	0004											—8					
269	9788380	0004							39								—7	
276	9784756	0020																
307	9710093	0006						3										
322	9784749	0024		—28														
338	9710093	0015										0						
352	9733763	0025										12						
369	9733776	0072							6									
384	9805657	0010											43					
403	9805656	0008								42								
407	9787132	0004											—28					
437	9785604	0018							—3									
473	9788382	0003															—2	
493	9806899	0006											—10					
505	9710281	0018									2							
509	332584	0054			15													
526	9787132	0012						39										
544	9787133	0010								40								

Job no.	*Code no.*	*Quantity*	*Shop stations*															*Split*
			1	*2*	*3*	*4*	*5*	*6*	*7*	*8*	*9*	*10*	*11*	*12*	*13*	*14*	*15*	
554	9788381	0002															−2	
555	9806457	0030										−8						
558	9789509	0003										19						

Table 5
Inquiry Report

Job no.	*Previous*		*Job activity*	*Shop stations*															*Split*
	Station	*Quantity*		*1*	*2*	*3*	*4*	*5*	*6*	*7*	*8*	*9*	*10*	*11*	*12*	*13*	*14*	*15*	
022																			
022	05	0150	–	–	–	–	–	–	×	×	–	×	×	×	–	–	×	×	
089																			
089	11	0066	×	–	–	–	–	–	–	–	–	–	–	–	×	×	–	×	
187																			
187	14	0096	×	–	–	–	–	–	–	–	–	–	–	–	–	–	–	×	S
556																			
556	07	0196	×	–	–	–	–	–	–	–	–	×	×	–	×	–	×	×	

information accumulated for the month is one of the inputs to the Customer Load Report.

Special features of the system

Certain features of the production control system warrant high lighting either because of the central role they play in the successful operation of the system or because of their intrinsic value as novel and interesting features.

The capability of on-line status reporting. The system can be queried in connexion with the status of the jobs in the shop. This 'on-line' status inquiry feature is restricted in two respects: one can obtain only *partial* information on the jobs, and only during certain periods of time. Thus, one can obtain the *quantity surviving* as well as the *future route* of any lot; however, one cannot obtain its dynamic priority or the quantity scrapped. The period of time during which this feature is available is limited by the demand on the computer by the various components of the system. Under one-shift operation, this period is reduced to approximately three hours during the eight-hour shift. The period is obviously expanded if second-shift operation on the computer is utilized.

An example of the inquiry report is given in Table 5. After loading the appropriate programs, the operator types the job number about which he wishes to inquire; say job 022. The computer prints out the status information on the following line. It repeats the job number, followed by: the 'last station' heard from (station 5), the last quantity surviving (150 units), followed by Xs if the job is routed to a station, and dashes if it is not (either because it has already passed through the station or because it was not designated to be worked on in that station). The column with the heading 'Job Activity' will contain a dash if *no activity* has been reported on the job since the previous updating of the master tape and will contain an 'X' if the job has moved since that updating. Thus job 022 was last reported from station 5, and has *not* moved since; while jobs 089, 187, and 556 did move from their previously-reported stations to stations 11, 14 and 7, respectively.

The treatment of 'urgent' jobs. Urgent jobs constitute an insignicant fraction of the actual number of jobs in the shop. However, their importance is completely out of proportion to their number. They are almost 'hand-carried' from one operation to the next, with all levels of supervision and management taking special interest in their progress. Any 'Urgent' job may exist in the shop for only a fraction of the normal status reporting cycle (which is one half week). Therefore, it was necessary to provide the capability of monitoring these jobs by the computer system and thus relieving supervision from this task.

'Urgent' jobs are not included in the master tape, hence no manufacturing data are available except the quantity surviving and the route. The extremely high dispatching priority of these few jobs causes them to move fast in the shop – in fact, their average manufacturing time is a small fraction of the 'normal' manufacturing interval, and closely approximates their production time (that is, almost no delay).

As an 'Urgent' job moves through the shop, it is reported in the usual fashion. Inquiry concerning the status of any of these 'Urgent' jobs is of course possible through the 'on-line' status reporting feature discussed previously.

The determination of the need for process capability studies. The Job Status Report, Table 3, serves a second and most important function other than presenting the complete status of each job in the shop. To wit, it provides the engineering department with the data necessary for the evaluation of the process capabilities at each of the fifteen reporting stations.

As can be seen, the body of the report contains the quantity completed at each station. Excessive 'dropout' at any station can be easily gleaned from such a report, and action to investigate and ultimately remedy the cause of such defective product can be undertaken. Historical data on any code as it progressed through the various operations can be easily compiled, which leads to accurate and realistic estimates of the expected 'shrinkage' in any lot. This, in turn, leads to more realistic loading and scheduling estimates.

Built-in error detection. As each transmission is entered in the memory the computer functions as a watchdog on certain errors. The new quantity is compared with that in the memory, and a typewriter printout is generated if the new quantity exceeds the old. Similarly if the station passed is not correct according to the prespecified route of the job, a printout is also generated. The printouts indicate the nature of the error so that immediate action may be taken.

The importance of these error-detection techniques transcends the need for an error-free status report. It affects the attitudes as well as the performance of the individuals responsible for transmitting the information through the remote input devices. The continuing occurrence of certain errors emanating from a specific reporting station leads to the identification of the individuals responsible for the errors and their further instruction.

Insurance against catastrophic breakdowns. The system was designed to recover quickly from any breakdown in equipment or personnel. The error-detecting features outlined are examples of recovery from personnel error. Several features have been included to insure fast recovery from equipment failure.

For example:

1. The shop transmissions are punched on paper tape prior to being processed by the computer. This insures continuing acceptance of shop transmissions even in the case of computer failure. It also insures the possibility of reconstructing the memory of the computer from the transmissions of previous days.

2. The availability of a spare master input device with a dial switch that enables entering transmission relative to any of the fifteen shop stations insures the ability to transmit in case of failure of one or more input devices.

3. The complete failure of the remote data transmission system still does not prevent access to the computer, which is programmed to accept simulated transmissions from the typewriter.

4. The Teletype page printer acts as a standby in the case of the failure of the off-line typing unit and vice versa.

5. Finally, there is a manual 'emergency' data collection procedure available in the case of complete failure of all components of the system. The latter can always be restarted from the latest master tape.

Evaluation and Conclusions

The system outlined in the previous sections has been in operation for several months. This enables us to evaluate, in an objective manner, the mathematical models as well as the various procedures that constitute the control system.

It is evident to us that the feedback, or status reporting, portion of the system is the most straightforward and readily-accepted portion of the whole system. This is due to several important factors, not the least of which is the fact that status reporting is an old and familiar activity in all shops. The contention that now the reports are generated by a computerized data-transmission-and-computation system is fundamentally irrelevant.

The various reports that relay the status to the shop management proved to be highly successful. In particular, the Shop Station Report is being utilized for the purposes which were not even contemplated in the design phase of the system. Now, this report is used to *forecast* the manpower requirements in the shop as a whole as well as for some particular stations. The average daily output as well as the variability in output of each station are computed by the shop first level supervision. These figures serve as a control mechanism in a similar fashion to the well-known statistical quality control charts. A simple extension of these figures, compiled with management's knowledge of the expected 'product mix' in future periods, yields reasonably accurate forecasts of future required manpower and equipment.

The *dynamic updating* of the priorities according to the progress of the jobs in the shop is meaningful and useful to the operating personnel. It is interesting to note that the existence of a physical meaning to the priorities rendered them understandable to the shop, and consequently a 'practical' scheme to be closely followed. In this respect, the control system proved to be educational as well as regulatory.

The monthly shop loading with economic manufacturing quantities was a new concept to the shop, and hence its acceptance was slow. Almost a year had to pass (between the first discussion of the concept and its final implementation) before the few ideas embodied in the procedure were accepted as 'practical' by the first and second levels of shop supervision. We believe that 'speeding up' the phases of indoctrination, demonstration and final application is not only useless but may be harmful. In many instances the new procedures seem, on the surface, to run counter to the direct interests of the operating personnel, in particular with regards to their earnings and reported 'efficiency'. Time is required to dispel such fears, as well as to educate them in the use of the new tools.

The various safeguards that are *built in to the system* against human error, equipment error, and catastrophic breakdowns are of the utmost importance. They are not only a safeguard for efficient system operation, but a necessary component for the building up of shop confidence in the new tool placed at their service. It is not possible for us to measure the psychological impact on the transmitting operator of his knowledge that 'he cannot fool the system for a long period of time', and that sooner or later his reports and his identification will be singled out.

The system has provided the shop management with a powerful tool for the control of production. Considerable economic savings were accomplished in both direct labor hours (for example in the elimination of end-of-month in-waiting inventory count) as well as indirect labor (for example, dispatchers, clerks, etc.). Several tangible though not measurable economic benefits were realized through the reduction in the total manufacturing interval, reduction of in-waiting inventories, and better customer relations. The system realizes the principle of placing the tool of control in the hands of the person (or persons) responsible for achieving that control. This is diametrically opposed to the concept of centralized time-shared computing and data-processing devices, which is the rule in manufacturing plants. The performance of the system is a proof of the validity of the concept, which is one step in the direction of de-centralization of the functions of manufacturing control.

References

BOWMAN, E. H. (1959), 'The schedule-sequencing problem', *Operations Research*, vol. 7, pp. 621–4.

CANNING, F. L. (1959), 'Estimating load requirements in a job shop', *J. Indust. Eng.*, vol. 10, pp. 447–9.

DEININGER, R. L. (1960), 'Human factors engineering studies of the design and the use of pushbutton telephone sets', *Bell System Technical Journal*, vol. 39, pp. 995–1012.

DZIELINSKI, B. P., BAKER, C. T., and MANNE, A. S. (1961), 'Simulation tests of lot size programming', paper presented to the Factory Scheduling Conference, Graduate School of Industrial Administration, Carnegie Institute of Technology.

MAGEE, J. F. (1958), *Production Planning and Inventory Control*, McGraw-Hill, p. 203.

MANNE, A. S. (1958), 'Programming of economic lot sizes', *Management Science*, vol. 4, pp. 115–35.

MANNE, A. S. (1960), 'On the job-shop scheduling problem', *Operations Research*, vol. 8, pp. 219–23.

ROGERS, J. (1958), 'A computational approach to the economic lot scheduling problem', *Management Science*, vol. 4, pp. 264–91.

SANDMAN, P. (1961), 'Empirical design of priority waiting times for jobbing shop control', *Operations Research*, vol. 9, pp. 446–55.

WAGNER, H. M. and WHITIN, T. M. (1958), 'Dynamic version of the economic lot size model', *Management Science*, vol. 5, pp. 89–96.

WAGNER, H. M. (1959), 'An integer linear programming model for machine scheduling', *Naval Res. Logistics Q.*, vol. 6, pp. 131–40.

12 E. Feldman, F. A. Lehrer and T. L. Ray

Warehouse Location Under Continuous Economies of Scale

E. Feldman, F. A. Lehrer and T. L. Ray, 'Warehouse location under continuous economies of scale', *Management Science*, vol. 12, 1966, pp. 670–84.

Warehouse Location is a nonconvex programming problem involving the geographic placing and sizing of intermediate facilities in distribution studies. The nonconvexities are caused by economies of scale associated with the cost of building and operating the facilities. A heuristic program has been developed for solving warehouse location problems when these economies are representable by continuous concave functions. The paper discusses the heuristics used and computational experience with the program on 'practical' problems. On the basis of two numerical examples for which an optimal solution was obtained through a special purpose experimental mixed integer programming code, it is conjectured (1) that near optimal solutions can be achieved using the heuristic program and (2) that optimal sizing and locating of facilities are very sensitive to the shapes of the warehousing cost functions.

Introduction

The warehouse location problem involves the determination of the number and sizes of service centers (warehouses) to supply a set of demand centers. The objective is to locate and size the warehouses and determine which demand centers are supplied from which warehouses so as to minimize total distribution cost. This distribution cost is the total transportation cost, which is assumed linear, plus the cost of building and operating the warehouses.

In most formulations of this problem the number of potential warehouse sites is finite, although Cooper (1963) has considered an infinite site case. In the finite case the problem cannot usually be considered as a transportation problem because of the 'economies of scale' associated with warehousing cost. Because big warehouses are more 'efficient' than small ones, total warehousing costs will rise as the number of warehouses is increased. Of course shipping cost falls as the number of warehouses is increased. Thus, the problem of minimizing total cost is to balance shipping against warehouse cost.

Balinski (1964), Kuehn and Hamburger (1963), and Manne (1964) have represented these economies by fixed costs, i.e., they associate a charge with 'opening' each warehouse. In these formulations the problem becomes one of discrete programming. These formulations are not satisfactory for those problems where the economies of scale affect warehousing cost over the entire range of warehouse sizes. It is this situation we will discuss in this paper.

Mathematically we will consider single-product problems of the form:

$$\text{Min} \sum_{ij} C_{ij} X_{ij} + \sum f_i(T_i)$$

such that

$$\sum_{i} X_{ij} = D_j \qquad X_{ij} \geqslant 0$$

where

$$\begin{aligned} X_{ij} &= \text{Flow from warehouse } i \text{ to demand center } j \\ T_i &= \sum_j X_{ij} = \text{Throughput of warehouse } i \\ C_{ij} &= \text{Unit cost of flow } X_{ij} \\ D_j &= \text{Demand at center } j \\ f_i &= \text{Warehousing cost function for warehouse } i. \end{aligned}$$

It is assumed that the f_i are continuous and concave over the range of interest. This is important from a computational standpoint and reasonable from an economic one.[1] It can be shown that concavity of the f_i and the absence of constraints on warehouse size give a computationally useful single assignment property, i.e., in the optimal solution no demand center will receive flows from more than one warehouse.

1. If the f_i were convex, the problem could be solved as a linear program after first approximating the f_i by piecewise linear functions. The concavity of the curves brought about by the 'economies of scale' lead to multiple-optima, and thus problems like these are not susceptible to conventional mathematical programming techniques.

The two approaches of Balinski and Kuehn-Hamburger are similar in that sets of warehouses are examined sequentially, with the set to be examined next determined by the costs associated with the previous sets. Here the similarity ends. Balinski showed a way to solve the problem using an integer programming algorithm[2] while Kuehn and Hamburger select the sets of warehouses using heuristics.[3]

In their formulations

$$f_i(T_i) = \begin{cases} b_i T_i + a_i, \quad a_i, b_i \text{ constants} & T_i \neq 0 \\ 0 & \text{otherwise.} \end{cases}$$

This representation of warehousing cost has the computational advantage that once a set of warehouses has been decided upon, the optimal shipments (flows) can be determined by solving a trivial linear program. Crucial use of this advantage is made in both formulations.

Extensions of Kuehn and Hamburger's Work

Our method of solving warehouse location problems follows Kuehn and Hamburger in that it uses heuristics, but we have extended it to handle concave f_i (see Figure 1). This extension consisted of developing heuristics for assigning customers to those warehouses that have been 'opened'. This assignment is no longer trivial for nonlinear f_i as it is in the linear case.[4]

2. We know of no computational experience with this algorithm. See Balinski (1964).

3. See Kuehn and Hamburger (1963) for a discussion of the nature of heuristic programming and an extensive bibliography on warehouse location problems.

4. Theoretically, any warehouse location problem with piecewise linear warehousing cost curves can be given a fixed cost formulation. A warehouse can be associated with each linear segment, the fixed cost being determined by extending the segment until it intersects the cost axis. The difficulty with this approach is computational. If each warehouse has a cost curve describable by five linear segments, then there would be five times as many warehouses in the fixed cost formulation. Our experience with the methods for solving warehouse problems discussed in the paper is that incremental computer time increases more rapidly with the number of warehouses than with the number of linear segments.

We developed a computer code for this extended method and used it for extensive testing on a number of 'practical' problems. The extension did not work well on all problems tested (the reasons for this are discussed below).

For this reason a different approach, based upon more 'reasonable' heuristics was developed. Both approaches have been

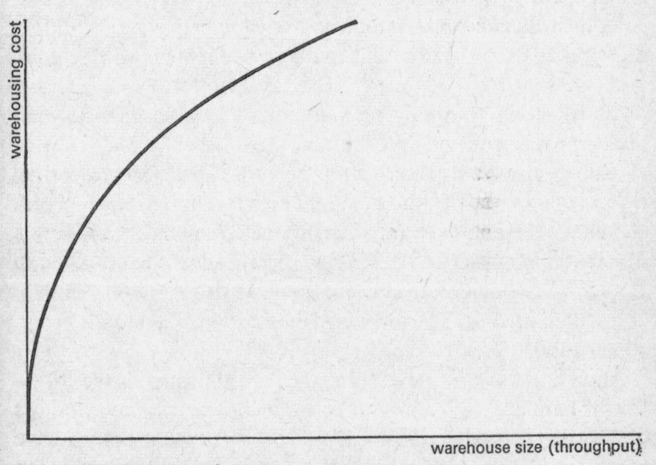

Figure 1 The warehousing cost curve

incorporated into a code for the I.B.M. 7094, used to solve several problems of interest. Some computational experience we have had with the code plus some of its capabilities are described below.

The important question of how well a heuristic program really works always arises. An experimental integer program was used to 'verify' the heuristic results. On a practical-size problem, comparison of the optimum solutions produced by the integer program with those produced by the heuristic program is given at the end of the text. On the whole the results obtained with the heuristic program have been most satisfactory.

A capability which we wished to incorporate is the allowance for different regional warehousing costs. For it is possible that an installation located some distance from any demand centers would be associated with low warehousing cost that would more than compensate for the high transportation costs involved. For this reason we desired a solution technique permitting the use of a different cost curve for every potential warehouse site. Our present technique fulfills this requirement and is very flexible with regard to the precise forms of the curves specified.

Another area in which we felt that modifications might be required was the method used to generate successively better solutions. Kuehn and Hamburger rely on sequential addition of warehouses, assuming that the best N suppliers will be contained among the best $N + 1$. They do mention, however, that *dropping* rather than adding warehouses is a possible way of attacking the problem.[5] We felt that there were two reasons why this drop approach would often be preferable. One reason is that the question of which warehouses to shut down or consolidate is the one most often facing management. Rarely is the distributer interested in designing a network from the ground up. Hence, we have developed in our code a 'drop' routine as well as an 'add' routine. The former is often better suited to the problem at hand.

Another reason for developing the 'drop' routine is the problem of handling *infeasible routes*, or routes for which no transportation costs were specified. We have worked with several actual problems in which the user was unwilling to even consider making deliveries over distances greater than some cut-off level. The explanation generally given for this decision was that very long routes must be debited with high-cost delay penalties, or in some cases total loss of sales. These infeasibility conditions are easily reflected in the formulation of the problem by using effectively infinite costs for transportation over the disallowed routes. Difficulty occurs, however, when the density of the transportation matrix falls so low as to make a three or four warehouse

5. This is to be expected. Heuristic programming by its very nature must be specialized to the particular 'type' of problem at hand. Sets of heuristics that will work well on some types of location problems may not work well on others. It is pointed out that by 'work well' we are referring to the problem of cost. Different heuristics might give different locations and sizes of warehouses but almost identical costs. The 'drop' approach was developed primarily because of cost considerations. In many cases the 'drop' approach gave solutions that were not only different in size and location from that given by the 'add' method, but also significantly lower in total cost.

solution impossible. The 'add' routine, which begins with the best single warehouse, will then attempt to get *feasible* as quickly as possible, a goal by no means necessarily coincident with that of getting optimal. We can, of course, supply a 'start solution' which will avoid this pitfall; such specification may, however, be tantamount to dictating the answer ultimately generated. By using the 'drop' approach, on the other hand, and starting with all or most of the potential sites in use, we can avoid this problem completely.

Heuristics for Handling Non-Linear Warehousing Costs

In order to effectively consider warehousing costs in solving the warehouse location problem as a whole, it is necessary to consider their effect on each of the individual warehouse-to-customer allocations determining the overall solution. The method by which the program deals with this problem is quite straightforward, and has proved to be very effective in practice. The approach taken is to evaluate, for each customer, the total incremental cost, including transportation and operating components, associated with shipments from each of the potential suppliers. For the problem we are considering the transportation aspects are trivial; unit costs are presumed independent of a shipment size. Incremental warehousing costs, on the other hand, are usually a decreasing function of warehouse throughput.

The method for assigning customers to warehouses, based upon both transportation cost and warehouse cost requires the determination of some 'reference level' for warehouse size. For the purpose of providing these required reference levels we define each warehouse's *local customer set* (LCS) as consisting of those customers to whom the warehouse is closest on a pure transportation cost basis. The warehouse can then be said to have a *local volume* that is the sum of the individual demands in its LCS. The quantity thus determined is taken as a preliminary measure of the extent to which a warehouse is centrally located, and is used to get an idea of which portion of the cost curve is most relevant to decisions involving this warehouse. There is, of course, no theoretical justification for the use of local demand as just defined; we have found, however, that it performs as well computationally as any other measures yet suggested. Since a widely applicable heuristic program must be flexible enough to handle problems of differing cost structures, we have left the definition of local customer setup to the user. He may experimentally vary LCS size, allowing each customer to pick the two nearest warehouses, three nearest, etc. Each step taken in this direction tends to reduce the program bias against isolated suppliers. Of course, the ultimate test of the various trials is in the quality of the solutions generated.

Once local volumes have been established, warehouse-customer allocation is independent of those made for all other customers. We simply examine the incremental cost of supplying a given customer from each of the available warehouses, assuming that these warehouses have, independent of the allocation in question, throughput levels equal to their local volumes.

The 'Drop' Approach

The 'drop' approach consists of starting with all (or most) potential warehouse sites being utilized as suppliers and eliminating them one at a time until no further cost reductions are realized. It is obvious that initial infeasibilities can be easily circumvented in this manner. There are several ways of choosing a reasonable start solution, the simplest of which is determined by minimization of transportation costs alone. We have found, however, that much better initial patterns can be obtained by using the local volume concept and considering operating costs when making initial assignments.

Although it is helpful to use local volumes initially, one great advantage of the 'drop' approach is that they are not needed in subsequent stages of analysis. The reader will recall that the principal purpose of a local volume is to provide a very rough indication of the ultimate size of a warehouse, given that it will be in the solution. This crude measure is about the best we have found for use in the 'add' program. When dropping warehouses, however, we have available at each stage of the solution process a far better set of indicators of ultimate warehouse size. These are simply the actual throughputs characterizing the pattern in use at that time. We are able, therefore, to make fuller use of the solution at each stage in working towards the final solution.

The actual throughputs associated with the start solution will presumably serve as better 'reference points' in computations of incremental operating costs than will local volumes. Before proceeding to the next stage of analysis, we make use of these throughputs in the present stage – computing a new solution based on these 'latest' reference points. The new solution will provide still another set of volumes which will be used similarly. This 'cycling' procedure continues until no further improvement

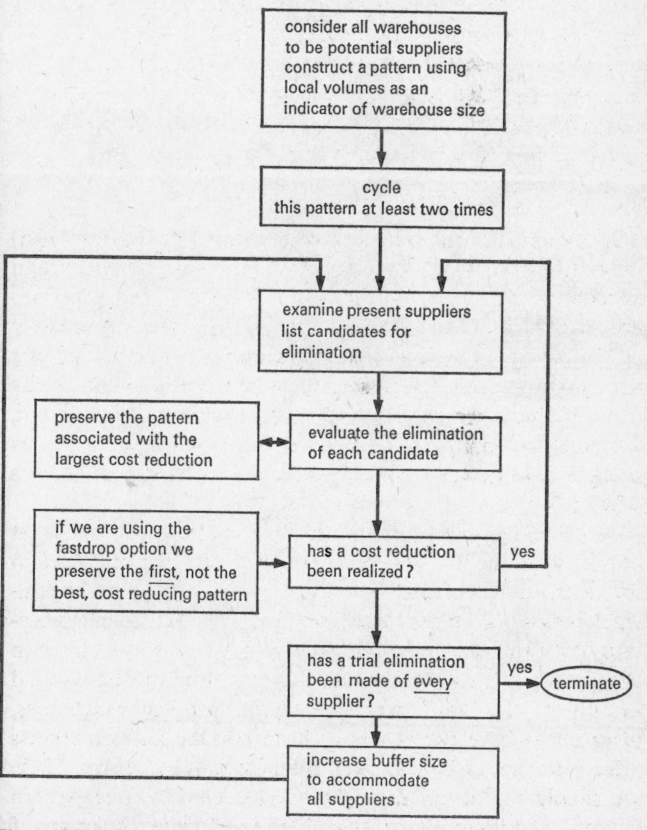

Figure 2 The drop approach

Table 1

Customer location	Potential warehouse site	Factory site	Customer location	Potential warehouse site	Factory site
New York, N.Y.	×		Fort Worth, Texas		
Chicago, Ill.	×		Long Beach, Calif.		
Los Angeles, Calif.	×		Birmingham, Ala.	×	
Philadelphia, Pa.	×		Oklahoma City, Okla.	×	
Detroit, Mich.	×		Rochester, N.Y.		
Baltimore, Md.			Toledo, Ohio		
Houston, Texas	×		St. Paul, Minn.		
Cleveland, Ohio	×		Norfolk, Va.	×	
Washington, D.C.	×		Omaha, Nebr.	×	
St. Louis, Mo.	×		Fall River, Mass.		
Milwaukee, Wis.	×	×	Miami, Fla.	×	
San Francisco, Calif.	×	×	Akron, Ohio		
Boston, Mass.	×		El Paso, Texas	×	
Dallas, Texas	×		Jersey City, N.J.		
New Orleans, La.	×		Tampa, Fla.		
Pittsburgh, Pa.	×		Dayton, Ohio		
San Antonio, Texas	×	×	Tulsa, Okla.		
San Diego, Calif.	×		Wichita, Kansas	×	
Seattle, Wash.	×		Richmond, Va.		
Buffalo, N.Y.	×		Syracuse, N.Y.		
Cincinnati, Ohio	×		Tucson, Arizona		
Memphis, Tenn.	×		Des Moines, Iowa		
Denver, Colo.	×		Providence, R.I.		
Atlanta, Ga.	×		San Jose, Calif.		
Minneapolis, Minn.	×		Mobile, Ala.		
Indianapolis, Ind.			Charlotte, N.C.	×	×
Kansas City, Mo.	×		Alburquerque, N.M.	×	
Columbus, Ohio			Jacksonville, Fla.	×	
Phoenix, Arizona	×		Flint, Mich.		
Newark, N.J.			Sacramento, Calif.		
Louisville, Ky.			Yonkers, N.Y.		
Portland, Ore.	×		Salt Lake City, Utah.	×	
Oakland, Calif.			Worcester, Mass.		
Austin, Texas			Topeka, Kansas		
Spokane, Wash.	×		Glendale, Calif.		

Customer location	Potential warehouse site	Factory site	Customer location	Potential warehouse site	Factory site
St. Petersburg, Fla.			Beaumont, Texas		
Gary, Ind.			Camden, N.J.		
Grand Rapids, Mich.			Columbus, Ga.		
Springfield, Mass.			Pasadena, Calif.		
Nashville, Tenn.			Portsmouth, Va.		
Corpus Christi, Texas			Trenton, N.J.		
Youngstown, Ohio			Newport News, Va.		
Shreveport, La.			Canton, Ohio		
Hartford, Conn.			Dearborn, Mich.		
Fort Wayne, Ind.			Knoxville, Tenn.		
Bridgeport, Conn.			Hammond, Ind.		
Baton Rouge, La.			Scranton, Pa.		
New Haven, Conn.			Berkeley, Calif.		
Savannah, Ga.			Winston-Salem, N.C.		
Tacoma, Wash.			Allentown, Pa.		
Jackson, Miss.			Little Rock, Ark.	×	
Paterson, N.J.			Lansing, Mich.		
Evansville, Ind.			Cambridge, Mass.		
Erie, Pa.			Elizabeth, N.J.		
Amarillo, Texas	×		Waterbury, Conn.		
Montgomery, Ala.			Duluth, Minn.		
Fresno, Calif.			Anaheim, Calif.		
South Bend, Ind.			Peoria, Ill.		
Chattanooga, Tenn.			New Bedford, Mass.		
Albany, N.Y.			Niagara Falls, N.Y.		
Lubbock, Texas			Wichita Falls, Texas		
Lincoln, Nebr.			Torrance, Calif.		
Madison, Wis.			Utica, N.Y.		
Rockford, Ill.			Santa Ana, Calif.		
Kansas City, Kansas			Saninaw, Mich.		
Greensboro, N.C.			Reading, Pa.		
Waco, Texas			North Platte, Nebr.		
Columbia, S.C.			Salina, Kansas		
Raleigh, N.C.			Grand Junction, Colo.		
Cedar Rapids, Iowa			Boise, Idaho	×	
Pueblo, Colo.			Idaho Falls, Idaho		

Customer location	Potential warehouse site	Factory site	Customer location	Potential warehouse site	Factory site
Abilene, Texas			Lewiston, Idaho		
Sioux City, Iowa			Logan, Utah		
Orlando, Fla.			Cedar City, Utah		
Charleston, W. Va.			Roswell, N.M.		
Huntingdon, W. Va.			Las Cruces, N.M.		
Portland, Me.	×		Yuma, Arizona		
Bangor, Me.			Flagstaff, Ariz.		
Concord, N. H.			Las Vegas, Nevada	×	
Harrisburg, Pa.			Reno, Nevada	×	
Roanoke, Va.			Eugene, Oregon		
Tallahassee, Fla.			Laredo, Texas		
Sault Ste. Harte, Mich.			Lawton, Okla.		
Springfield, Mo.			Fort Smith, Ark.		
LaCrosse, Wis.			Presque Isle, Me.		
Fargo, N. Dak.			Burlington, Vermont		
Grand Forks, N. Dak.			Twin Falls, Idaho		
Minot, N. Dak.			Williston, N. Dak.		
Bismarck, N. Dak.	×		Paducah, Ky.		
Sioux Falls, S. Dak.			Odessa, Texas		
Rapid City, S. Dak.			Medford, Oregon		
Aberdeen, S. Dak.			Liberal, Kansas		
Billings, Mont.	×		Pierre, S. Dak.	×	
Butte, Mont.			Eureka, Calif.		
Great Falls, Mont.			Green Bay, Wis.		
Missoula, Mont.			Augusta, Ga.		
Cheyenne, Wyo.			Gallup, N.H.		
Casper, Wyo.	×		Pendleton, Ore.		
Sheridan, Wyo.			Elko, Nevada		
Grand Island, Nebr.			Del Rio, Texas		

in total cost is observed, at which time we begin making trial eliminations. Note that we are not guaranteed that all warehouses with which we start cycling will be in the solution when we are through. The fact that suppliers can be eliminated in this manner tends to reduce the computer time necessary to reach the final solution.

We originally expected that use of the drop approach would, in addition to its other advantages, facilitate the 'preliminary' or 'buffer' evaluation stage of the location process.[6] Buffer evaluations in the add routine proved to be as time consuming as detailed evaluations, and were finally eliminated. We thought, however, that warehouse size at any stage of an elimination process would quickly provide us with a good idea of which

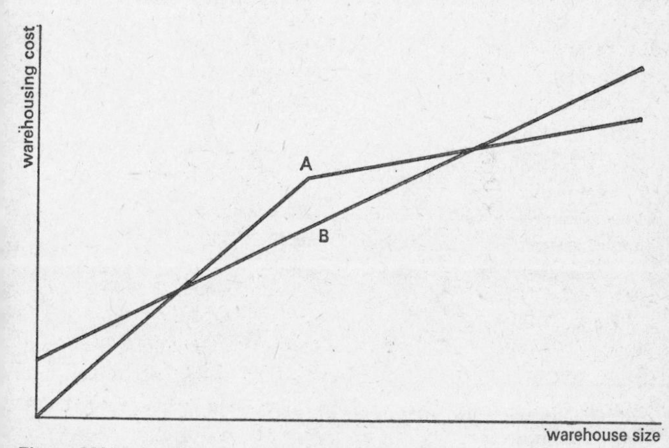

Figure 3 Warehousing cost curves used

suppliers should be dropped next, the smaller warehouses being the prime candidates. Experience has not borne out these expectations; we have seen cases in which the elimination of a large warehouse leads directly to an optimal solution, with all small warehouses remaining in the final patterns. This problem can be handled by using large buffer sizes to essentially ignore the buffer. We have also experimented successfully with a modified elimination technique: instead of dropping the warehouse yielding the biggest saving, drop the smallest one yielding *any* saving. Doubtless many examples could be found for which either technique would be inadequate either in terms of computing time or the quality of solutions produced; but our actual experience does suggest that the methods presently in use are capable of efficiently generating optimal or near optimal solu-

6. Buffer is used the same as in Kuehn and Hamburger (1963).

tions to a broad class of problems. A flow chart of the 'drop' program appears in Figure 2.

Use of the Program

The program was tested initially on the problems most readily available, those formulated and solved by Kuehn and Hamburger. No solutions generated were higher in cost than the Kuehn and Hamburger patterns, but no large improvements were observed.

Table 2
Cost Curve Data

Curve A		*Curve B*	
Volume	*Cost*	*Volume*	*Cost*
0	0	0	7140
5000	50,000		
15,000	100,000	10,000	71,440

These results were obtained by both the 'add' and 'drop' methods. Running times were under one minute per case on an I.B.M. 7094.[7] Because of the satisfactory running times experienced, all problems we consider are always solved using both the 'add' and 'drop' methods and the best answer is selected.

Testing the Code

Following these tests we decided to construct a much larger problem which would fully test the capabilities of our code. The approach taken was similar to that of Kuehn and Hamburger. Four factories, forty-nine warehouse sites, and 200 customer locations were laid out on a map of the United States. Populations of the cities were used to represent customer demands. Transportation costs were assumed to be linear functions of

7. Subsequent experience has indicated that realistic business problems, using 25 potential warehouse sites to service 150 customers, can be solved in 2–3 minutes.

distance, with each customer being feasible to the fifteen closest warehouses.[8] Table 1 is a list of the cities used and the functions performed at the various locations.

The program will handle warehousing cost curves consisting of any number of linear segments, allowing us to approximate a continuous concave curve with any desired degree of accuracy. For the purpose of illustration, however, we will confine our interest to a 'single-knee' curve such as A in Figure 3, as well as to B, a linear relationship with fixed cost component of the type used by Kuehn and Hamburger. There is, in fact, one straight line of particular interest for use as curve B, namely that which provides the best linear fit to the three points determining curve A. A comparison of solutions using first A and then B will provide some indication of the sensitivity of the problem to the mode of warehousing cost representation. The curves used are described numerically in Table 2.

Computational Experience

We attacked both Problem A and Problem B (corresponding to the two cost curves) with each of the heuristic techniques available. As we had no prior knowledge of the nature of the solution we used each approach in conjunction with the other, feeling that the two could be used as a 'package' far more effectively than could either one alone. Thus, for example, we eliminated distortions due to early infeasibilities in the 'add' approach by providing a start solution using the best warehouses in the 'drop' solutions. We were careful, however, not to *force* the two approaches to generate the same patterns.

Problem A. Problem A was examined first. The 'drop' program, produced the solution displayed in Figure 4; the 'add' run yielded the pattern in Figure 5. The total cost of the 'drop' solution was lower by about 3 per cent. As is obvious from

8. Demands were set equal to population, in thousands, as determined by the 1960 census. Transportation costs were handled in a less precise manner. Using an ordinary roadmap we measured approximate airline distances between cities. Factory-warehouse costs were then taken to be equal to the distances in hundreds of miles, while warehouse-customer routes had costs equal to twice the distance.

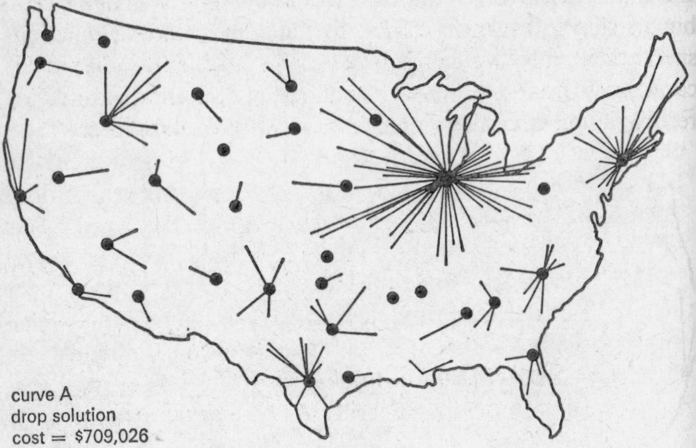

curve A
drop solution
cost = $709,026

Figure 4

examination the two patterns differ significantly, the 'add' pattern having slightly more than *half* as many warehouses as the 'drop'. It seems that during the early stages of sequential addition many warehouses were deemed uneconomical because they were in the vicinity of a huge supplier having a volume 'over

curve A
add solution
cost = $730,176

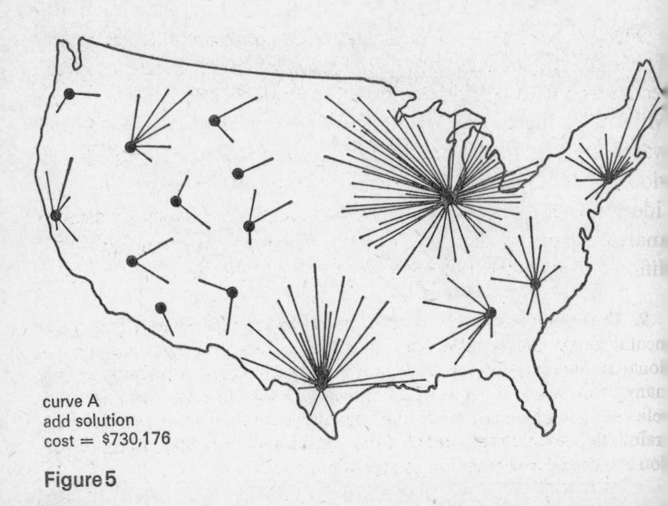

Figure 5

the knee' in curve A. Later on in the analysis the huge warehouses slowly lost their volume to candidates who were added
successfully, making addition of some previously discarded
candidates desirable. There was, however, no provision for
reconsidering discarded warehouses, so they were lost forever.

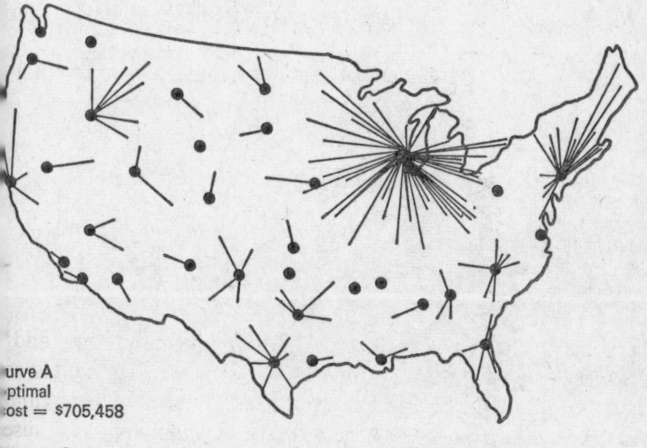

igure 6

The optimal solution to Problem A is illustrated in Figure 6.[9]
The pattern is quite similar to that produced by the 'drop'
echnique, but large warehouses appear in Philadelphia and
Milwaukee instead of New York and Chicago. The absence of
warehouses in the nation's two largest cities in a problem obviously dominated by transportation and population con-
iderations is an occurrence that we feel would escape any human
nalyst as surely as it escaped the heuristic program. The cost
difference was only 0·5 per cent, however, so the heuristic solu-

9. The optimum solution was obtained by use of a special purpose experi-
nental mixed integer code. This code was satisfactory, from a computa-
ional standpoint, for the problems discussed in the text. However, for
nany problems of interest (50 warehouses, 5 segments per cost curve) we
elieve the integer code would not be competitive with the heuristic pro-
ram. All heuristic runs preceded the optimization attempts, so no distor-
ion occurred due to prior knowledge of our goals.

tion provided a pattern that is quite acceptable from a practica standpoint.

It is obvious from examination of Figure 6 that knowledge o the solution would even influence problem formulation. A ware- house at Omaha was built to supply North Platte, Nebraska even though Omaha itself was served by Milwaukee. The reaso for this was that Milwaukee was not one of the fifteen supplier nearest to North Platte, so no route connecting the two wa allowed. As a result the nearest warehouse was utilized fo

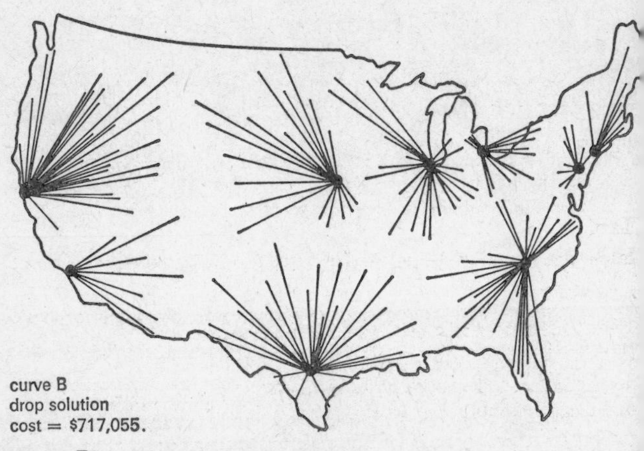

Figure 7

this single customer. One would naturally examine relaxatic of the cut-off criterion under such circumstances, but the pro lem as it stands is sufficient for our purposes. We must distingui between solutions of a *given* problem and proper *formulati* of that problem from a physical and economic standpoint.

Study of the Problem A cases shows that routine use of t program would not lead to satisfactory results and thus, the us must have insight for selecting starting solutions. The 'dro program was overly biased by the dominance of New Yo and Chicago in the pure transportation solution used as a sta Milwaukee and Philadelphia were quickly eliminated. The 'ad routine was also biased by early solutions in which New Yo

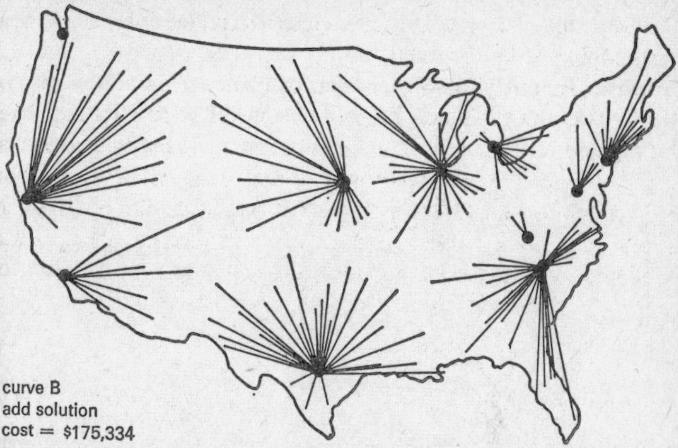

curve B
add solution
cost = $175,334

Figure 8

Table 3
Sensitivity to Warehousing Cost Assumptions

1. Heuristic pattern under curve B assumptions	Figure 8
2. Evaluation of (1) under curve A conditions	$776,122
3. Heuristic solution under curve A assumption	$709,026
4. Loss due to simplification (2–3)	$67,096

and Chicago were used. Once the two of them were employed as suppliers, Milwaukee and Philadelphia were soon discarded. Only a start solution specifying the latter could have forced them into the final answer. A second problem facing the 'add' was the fact that early solutions differed so greatly from the optimal solution that the sequential approach seemed to break down entirely. The patterns ultimately generated bore little resemblance to those being sought.

Problem B. The solutions to Problem B (Figures 7 and 8) were strikingly different from those just discussed. Both heuristic approaches eventually generated a nine-warehouse pattern. The only difference between the 'add' and 'drop' answers was that a warehouse at Philadelphia, rather than one at Washington, was employed by the latter. Total costs were virtually identical; however, both techniques performed satisfactorily on this problem, generating a much better 'cluster' of total costs for the cases examined than they did for Problem A. The spread between

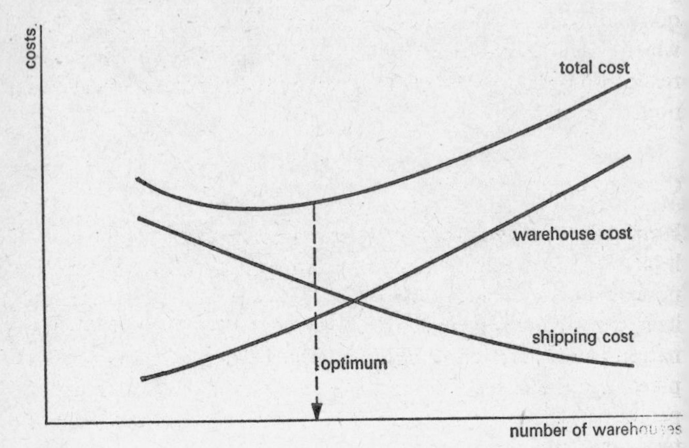

Figure 9

the best and worst cases was only 0·3 per cent compared to almost 3 per cent for the earlier problem. The *optimal solution* in this case is precisely the same as the 'add' pattern generated by the heuristic and displayed in Figure 8.

The Need for Handling Continuous Economies of Scale

If we assume that the 'single knee' curve A is the true representation of warehousing cost (it is more typical), and that only a 'fixed' cost program is available, then a simplifying curve such as B would have to be used to approximate A.

It is apparent that curves A and B lead to significantly differen warehouse-customer patterns. This is illustrated in Table 3, where an evaluation of the pattern corresponding to the latte assumption is made under cost conditions corresponding to the former.

The figures show that an additional cost of almost 10 per cen can result from simplification of the warehousing relationship and, thus, indicates the need for having the capability to hand treat continuous economies. Also, this points up the necessity of making use of *all* available cost information; and even more important, of searching for information of this type. The company that can put together a continuous nonconvex curve is quite unlikely to overlook simple fixed costs; however, a firm which tries to express the warehousing function as a straight line relationship may very well be ignoring the important effects of nonlinear behavior.

Conclusions

Heuristic techniques *can* generate near optimal solutions to large scale warehouse location problems having continuous nonconvex warehousing cost functions. The nature of the problem itself will dictate whether sequential addition or sequential elimination is the more appropriate approach. Each method has its place, but the latter appears to function over a wider range of problems. There may, however, be problems for which neither technique would be suited.

'Optimal' patterns are highly sensitive to the form of the warehousing cost curve. One should therefore avoid oversimplification during the formulation phase.

References

Balinski, M. L. (1964), 'On finding integer solution to linear programs', *Mathematica*.

Cooper, L. (1963), 'Location-allocation problems', *Operations Research*, vol. 11.

Kuehn, A. A., and Hamburger, M. J. (1963), 'A heuristic program for locating warehouses', *Management Science*, vol. 10.

Manne, A. S. (1964), 'Plant location under economies-of-scale – decentralization and computation', *Management Science*, vol. 11.

Part Four Theoretical Developments

This final part is intended mainly to provide a fuller account of some of the mathematical programming techniques encountered in the previous part. Wolfe's paper gives a theoretical treatment of many useful non-linear programming techniques besides those already encountered, and that of Charnes and Cooper gives a comprehensive account of chance constrained programming. Another form of probabilistic programming, linear programming under uncertainty, is introduced by the paper of Elmaghraby. The next paper, by Haley and Smith, is not closely related to any of the previous ones, but is important in so far as it illustrates the kind of technique that may be developed to deal with problems of a special structure. Finally, Beale's paper gives a splendid survey of the difficult field of integer programming, and that of Agin expands the notion of branch and bound, one of the many possible techniques for integer programming but also an important concept in its own right.

13 P. Wolfe

Methods of Non-Linear Programming

P. Wolfe, 'Methods of non-linear programming', in J. Abadie (ed.), *Non-Linear Programming*, North-Holland Publishing Co., 1967, pp. 99–131.

Preliminaries

1. *Notation*

In what follows, $x = (x_1, \ldots, x_n)$ and similar letters will refer to points or vectors in n-dimensional Euclidean space. The ordinary inner product will be written either xy or $x \,.\, y$, and the length of the vector x is $|x| = (x \,.\, x)^{\frac{1}{2}}$. The character θ will always denote a number between zero and one, and $\bar{\theta} = 1 - \theta$. Thus $\theta x + \bar{\theta} y$ is a point on the line segment joining x and y. A set of points is *convex* if $\theta x + \bar{\theta} y$ belongs to it whenever x and y do.

The functions $f(x) = f(x_1, \ldots, x_n)$, etc., with which we deal will always be continuous, defined over some closed and usually convex set S. We write $\nabla f(x) = (\partial f/\partial x_1, \ldots, \partial f/\partial x_n)$ when all the derivatives exist, and if all the second derivatives exist then $\nabla^2 f(x)$ is the Hessian of f, the n-order matrix whose entry in row i, column j is $\partial^2 f(x)/\partial x_i \partial x_j$.

The distinction between row and column vectors will generally be ignored, the appropriate character being decipherable from the context when necessary. The symbol ^{T} will be used for transposition of matrices. Braces $\{, \}$ enclose the members of a set, the expression $\{z: \ldots z \ldots\}$ denoting the set of all z such that $\ldots z \ldots$. The problem of maximizing the function of f over the domain S is thus laconically stated as $\max \{f(x): x \in S\}$.

2. *Convexity*

The notion of the convexity of a function plays a key role in several parts of mathematical programming. In the characterizations given here we assume that f is defined on the convex set S from which the points x and y are drawn.

The function f is *convex* if:

The set $\{(x, z): f(x) \leqslant z\}$ is convex. **1**

$f(\theta x + \bar{\theta} y) \leqslant \theta f(x) + \bar{\theta} f(y)$ for all $x, y \in S$ and $0 \leqslant \bar{\theta} \leqslant 1$. **2**

For any x in the interior of S there exists an n-vector u such that

$$u(y - x) \leqslant f(y) - f(x) \text{ for all } y \in S. \quad \mathbf{3}$$

$\nabla^2 f(x)$ is positive semidefinite for all x in S. **4**

The first three characterizations are all equivalent, and each can serve as a definition. The fourth is meaningful only if all second derivatives exist; then it is equivalent also. We will not prove the equivalences, but briefly discuss the relationship of the characterizations here and in section 6. For a thorough discussion of these relations Fenchel's book (1953) is invaluable.

The set of **1** above is just the set of all points lying on or above the graph of the function f; see Figure 1 for an example of a convex function of one variable. Taking two points $(x, f(x))$ and $(y, f(y))$

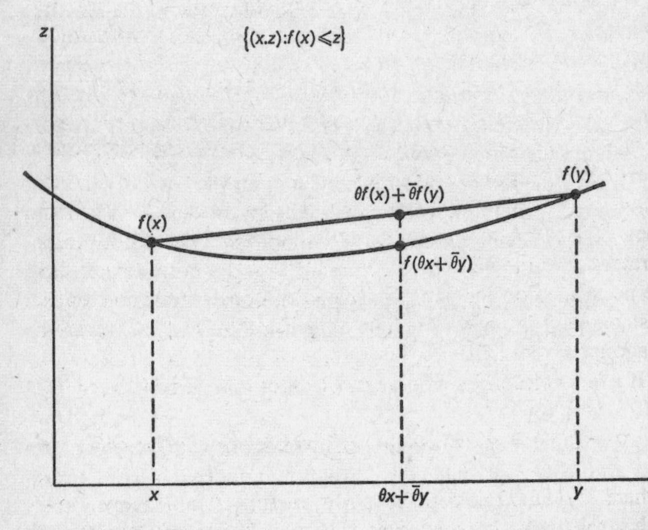

Figure 1 Convexity

on the graph of f, it is easy to see that **2** just expresses the convexity of that set. Characterization **2** states precisely that the function $f(\theta x + \bar{\theta} y)$ of the single variable θ shall be convex for every x and y, so that, by virtue of the equivalence, we can say that f is convex if it is convex on every line segment in S. Characterization **3** says that the set of **1** has a support at every point x, a general characteristic of convex sets. If f is differentiable at x then the vector u turns out to be $\nabla f(x)$, and the relation

$$\nabla f(x) \cdot (y - x) \leqslant f(y) - f(x) \qquad \mathbf{5}$$

is most useful. This relation and **4** are discussed further in section 7.

As a practical matter, the convexity of a mathematically given function is usually known through the formula given, through these rules of composition: Any linear function, or any convex function of a linear function, is convex; the sum of convex functions is convex. As one example, the sum of squares of linear functions is convex, and indeed a quadratic function of several variables is convex if and only if it can be written as such a sum.

It is sometimes convenient to check convexity by determining that $Q = \nabla^2 f(x)$ is positive semidefinite for all x. There are a few general rules that may be of assistance. If A is any matrix, then $A^T A$ is positive semidefinite; and so is a diagonal matrix with nonnegative entries. Q is positive semidefinite if and only if all its eigenvalues are nonnegative, for which the Hadamard-Gerschgorin theorem gives a sufficient condition: that $q_{ii} \geqslant \Sigma_{j \neq i} |q|_{ij}$ for all i. If these general rules cannot be applied, then checking Q may become tedious, although the procedure is straightforward: all that is required is that pivotal operations be performed on the matrix, using each nonzero diagonal entry just once. Q is positive semidefinite if and only if no negative diagonal entries appear.

If g is a twice-differentiable function of one variable, then

$$\nabla^2 g(f(x)) = g''(f(x))[\nabla f(x)]^T \nabla f(x) + g'(f(x)) \nabla^2 f(x) \qquad \mathbf{6}$$

where $\nabla f(x)$ is taken to be a row vector). If g is convex and either f is linear (in which case $\nabla^2 f(x) = 0$) or $g'(f(x)) \geqslant 0$ and

f is convex, then the matrix **6** is positive semidefinite. Thus a monotone increasing convex function of a convex function is convex, and in particular if f is convex then $-1/f$ is convex on the set $\{x : f(x) < 0\}$. (This is, however, easy to prove directly from **2**, too.)

A function is said to be concave if its negative is convex. The replacement of $f(x)$ by $-f(x)$ in **1–4** above leads immediately to a corresponding set of characterizations, the most commonly used of which is that f is concave if

$$f(\theta x + \bar{\theta} y) \geqslant \theta f(x) + \bar{\theta} f(y) \text{ for all suitable } x, y, \theta. \qquad \mathbf{7}$$

The role played by all these concepts in nonlinear programming will be mostly set forth in section 4.

3. *Quasiconvexity*

Quasiconvexity is a weaker property than convexity, but is sufficiently strong to have most of the desirable properties for mathematical programming that convexity does. This fact is of more theoretical interest than practical importance, because quasiconvexity is hard to test, and in practice functions which are quasiconvex but not convex seem to be rare.

Either of the equivalent propositions **8** or **9** serves as a definition of the quasiconvexity of the function f:

$$\{x : f(x) \leqslant a\} \text{ is convex for all } a. \qquad \mathbf{8}$$

$$f(\theta x + \bar{\theta} y) \leqslant \max \{f(x), f(y)\} \text{ for all } x, y \text{ in } S. \qquad \mathbf{9}$$

It is easy to see that characterizations **1** and **2** of convexity respectively imply **8** and **9**. Figure 2 sketches the function $f(x) = x^2/(1 + x^2)$, which is quasiconvex but not convex.

The function illustrated fails to be convex because its values 'do not rise fast enough' for large x. It is sometimes possible to apply a nonlinear change of scale to the values $y = f(x)$ of the function to make it convex; that is, to find a strictly monotone function g defined on the range of f so that $g(f(x))$ is convex. In the above example this is accomplished by taking $g(y) = 1/(1 - y)$. On the other hand the function $f(x_1, x_2) = x_2/x_1$, which is quasiconvex on any set in the interior of one of the four orthants, cannot be convexified in this way. The determinant of the Hessian of $g(f(x_1, x_2))$ for any twice-differentiable g (see 6 above) is $-[g'(f(x_1, x_2))]^2/x_1^4$, so that the Hessian is always indefinite.

Analagously to the definition of concavity, a function is *quasiconcave* if its negative is quasiconvex.

Unhappily, there seem to be no simple rules of composition for quasiconvex functions. Adding any linear term kx to the function of Figure 2 destroys its quasiconvexity, and so does

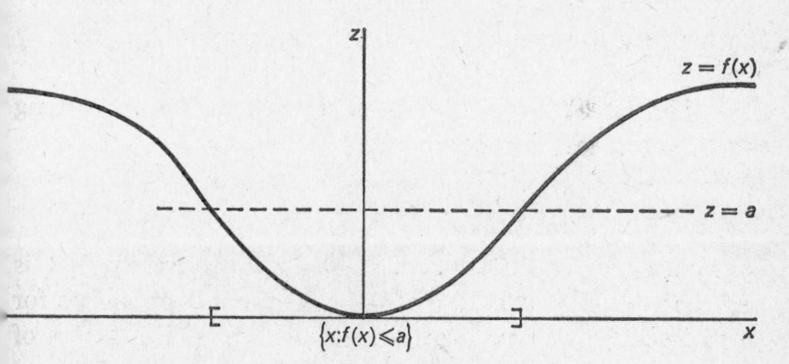

Figure 2 Quasiconvexity. $f(x) = x^2/(1+x^2)$

even adding an 'independent' linear term: $x^2/(1 + x^2) + y$, as a function of two variables has none of the desirable properties, as is shown by the fact that along the line $x = 3y$ it has three zeroes, between which it assumes both positive and negative values.

4. *Kinds of solutions*

The problem to which we are devoted is that of maximizing the function f on the set S. While the proper notion of 'solution' is the first of those given below, we cannot always be so demanding of the problem as to require it, and the succeeding weaker notions are sometimes all we can hope for.

$\bar{x}$ in S is a (*global*) solution if

$$f(\bar{x}) \geqslant f(x) \text{ for all } x \text{ in } S. \qquad 10$$

$\bar{x}$ in S is a *local solution* if for some $\delta < 0$

$$f(x) \geqslant f(x) \text{ for all } x \text{ in } S \text{ with } |x - \bar{x}| < \delta. \qquad 11$$

$\bar{x}$ in S is an *ε-solution* if

$$f(\bar{x}) \geqslant f(x) - \varepsilon \text{ for all } x \text{ in } S. \qquad \mathbf{12}$$

$\bar{x}$ in S is a *stationary point* if, for all y in S,

$$\limsup_{\theta \to 0+} \frac{f(\bar{\theta}x + \theta y) - f(x)}{\theta} \leqslant 0. \qquad \mathbf{13}$$

Obviously the ε-solution is necessary in a most general way, since a reasonable problem may not have a solution – say, that of maximizing $-1/x$ subject to $x \geqslant 1$. Further, no algorithm can terminate in the solution of the problem of maximizing $x - x^4$, since the 'solution' is not a rational number. We must then consider a problem completely solved if we have a finite process which can give an ε-solution for arbitrary positive ε. If, for some sequence $\varepsilon^k \to 0$ a sequence of ε^k-solutions x^k is obtained which converges to some x, then of course $\bar{x}$ is a global solution. It is important to note, however, that for most practical purposes the existence of such an $\bar{x}$ is not at all important; it is only the ability to choose a sufficiently good x^k that counts. In other words, the limiting behavior of $f(x^k)$ is the only thing of real significance.

All the procedures we will look at only provide local solutions of our mathematical programming problem. The knowledge of f employed by the procedures is based only on information regarding the immediate neighborhood of the best point found so far. If it is itself the best point of the neighborhood, it is a local solution, and we will generally have no clue as how to proceed further. This is an essential point in practice, and more attention should be paid to it than has been from the theoretical point of view. In some cases, to be discussed below, we may argue that any local solution will be global; but our knowledge of f is usually insufficient to permit that. In such a case we can only try to find all, or many, local solutions. Sometimes a subregion of S can be identified which is known to have only one local solution, so that S can be partitioned into regions in each of which a simpler search can be carried out. If we have obtained a sequence x^k of improved approximations to a local solution, we might suppose that no better solution could be found close to any x^k or perhaps even close to the convex hull of all of them, and thus delimit the remainder of S to be investigated. These matters seem very difficult to discuss theoretically, and form the heart of the art of extremizing difficult functions.

It is clear that any global solution is local, and that any local solution is a stationary point. By way of counterexample to the reverse propositions, the function $y = 8x^2 - 4x^3 - x^4$ has its only global maximum at $(-4, +128)$, has another local maximum at $(1,3)$, and another stationary point, which is not a local maximum, at $(0, 0)$.

The definition **13** of the stationary point requires that the directional derivative of f (if it exists) be nonpositive in any direction pointing toward some point y of S. If ∇f exists. **13** may be written

$$\nabla f(\bar{x})(y - x) \leqslant 0 \text{ for all } y \text{ in } S. \qquad \textbf{14}$$

If $\bar{x}$ is an interior point of S, so that points y and $\bar{x} - y$ may be chosen, then we must have $\nabla f(\bar{x}) = 0$, the classical case (but an uncommon one in our problem). A particularly valuable expression of **13**, the conditions of Kuhn and Tucker, will be developed in section 12.

The principal role that the notions of convexity and quasi-convexity play in mathematical programming is that of ensuring that the weaker types of solution, when found, are actually strong solutions. The theorems below, whose proofs are straight-forward, summarize the available facts:

Theorem: If f is concave, then any point having one of the three properties of being a stationary point, a local solution, or a global solution, also has the other two properties.

Theorem: If f is quasiconcave, then any strict local solution is a strict global solution.

(One of these solutions is called *strict* if the inequality of its definition **10** or **11** is replaced by a strict inequality, to hold for $x \neq \bar{x}$.)

As a counterexample to the deletion of 'strict' from the latter theorem the function defined by $y = (x + 1)x^2$ for $x < 0$, $y = 0$ for $x \geqslant 0$ has local solutions which are not global at each $\bar{x} > 0$, and incidentally a stationary point at $x = 0$ which is not a local solution.

5. *The nonlinear programming problem*

We will use this traditional statement of the nonlinear programming problem:

Given the continuous functions $g_1, \ldots, g_m$ and f, it is required to find the point x which maximizes f under the constraints $g_i(x) \geqslant 0$, $i = 1, \ldots, m$. We will assume that the set S satisfying all the constraints is convex. Since $S = \bigcap_i \{x : g_i(x) \leqslant 0\}$, S is convex if each g_i is quasiconvex.

If the variable x_{n+1} is added to the problem, we may restate it as requiring the maximization of the simple function x_{n+1} under the constraints above and the added constraint $x_{n+1} - f(x) \leqslant 0$. If f is concave then this new constraint function is convex, and our hypotheses are still good. Although if f is quasiconcave it does not follow that the constraint is quasiconvex, it is still true that any local solution is global.

In the case of linear constraints we write

$$g_i(x) = \sum_i a_{ij}\, x_j - b_i, \qquad i = 1, \ldots, m. \qquad 15$$

This case is of great interest because much of the work applicable to ordinary linear programming problems can be applied to it, and because problems with only mildly nonlinear constraints can sometimes be recast in this form.

In most of the subsequent work we shall suppose that a point satisfying the constraints of the problem is already at hand with which to begin the algorithm under study. In practical problems this is often the case. When it is not, it is generally possible to define a new problem, the so-called 'Phase One' problem, for which a starting point is readily at hand, and whose solution yields a starting point for the actual problem. (The term 'Phase One' is used in linear programming to denote that process of finding a first feasible solution, and 'Phase Two' to denote the subsequent solving of the whole problem.) There are several ways to define a function F whose maximum is zero if and only if the constraints can all be satisfied. The simplest seems to be

$$F(x) = -\sum_i \max\,(0, g_i(x)); \qquad 16$$

$F(x) = 0$ if and only if x is in S. This function is not, however, differentiable at $g_i(x) = 0$, which might be awkward, at least theoretically. But the function

$$F(x) = -\sum_i [\max(0, g_i(x))]^2 \qquad 17$$

is differentiable, and is, further, concave if the g_i are convex. Thus its maximum without constraints, solves the starting-point problem.

A variation on the use of **16** keeps the simplicity of the functions g_i and stays differentiable. Given x, define $I(\bar{x})$ as the set of all values of the index i for which $g_i(\bar{x}) > 0$, and define $F_x(x) = -\sum_{i \varepsilon I(\bar{x})} g_i(x)$. Note that $F_{\bar{x}}(\bar{x}) = F(\bar{x})$ as given by **16.** Now proceed with the solution of the problem

$$\text{Maximize } F_{\bar{x}}(x) \text{ subject to } g_i(x) \leqslant 0, \text{ all } i \;\varepsilon\; I(\bar{x}), \qquad 18$$

until a point x is found satisfying the above constraints for which some additional constraint $g_i(x) \leqslant 0$ is satisfied for some i in $I(\bar{x})$. Now let $\bar{x}$ be this new x, and return to **18**. In effect this process aims at satisfying all the constraints at once, retaining each as soon as it is satisfied. It is the standard procedure in good linear programming routines.

6. *The gradient and Taylor's series*

Following standard terminology, let us use o_i (the 'little o' function) to denote an arbitrary function, of scalars or vectors as necessary, with the property that

$$\lim_{x \to 0} \frac{o_i(x)}{|x|^i} = 0. \qquad 18$$

We say that a function f of several variables is differentiable at x if

$$f(x + \Delta x) = f(x) + \Delta f(x)\Delta x + o_1(\Delta x) \qquad 19$$

for all sufficiently small Δx. The existence of $\nabla f(x)$ alone is not enough to guarantee the differentiability of f, of course: the function

$$f(x, y) = \begin{cases} xy/(x^2 + y^2) & \text{if } (x, y) \neq 0 \\ 0 & \text{if } (x, y) = 0 \end{cases}$$

is not differentiable at the origin, but $\nabla f(x) = 0$ there.

For any x and any nonzero y the quantity

$$\lim_{t \to 0+} \frac{f(x + ty) - f(x)}{t} \tag{20}$$

is called the directional derivative of f at x in the direction y (it is sometimes convenient to require $|y| = 1$). If f is differentiable then the directional derivative exists, but the reverse is not always the case. The function $f(x_1, x_2) = (3x_1^2x_2 - x_2^3)/(x_1^2 + x_2^2)$ is differentiable everywhere except at the origin, but possesses a directional derivative there: in polar co-ordinates $f(x_1, x_2) = r \sin 3\theta$, and the directional derivative in the direction y is $|y| \sin 3\theta$; and this cannot be expressed as $u \,.\, y$ for any u.

As Fenchel (1953) shows, if f is convex then its directional derivatives exist in any direction y such that $x + ty \in S$ for some $t > 0$. It will be finite if x is interior to S, and might be only $-\infty$ otherwise.

We will say that f is twice-differentiable at x if

$$f(x + \Delta x) = f(x) + \nabla f(x)\Delta x + \tfrac{1}{2}\Delta x \nabla^2 f(x)\Delta x + o_2(\Delta x) \tag{21}$$

for all sufficiently small Δx. The third and higher-order terms of the Taylor's series expansion of f thus indicated by o_2 are uniformly ignored in mathematical programming algorithms, as in most of numerical analysis. They are hard to calculate, and we have already enough trouble dealing with second-order terms; indeed, many procedures ignore them, too.

Any set $\{x : f(x) = a\}$ is a *contour* of f. The linear manifold having highest contact with the contour is the tangent plane, whose equation is just $\nabla f(\bar{x})(x - \bar{x}) = 0$ if the contour passes through $\bar{x}$. The gradient $\nabla f(\bar{x})$ is then normal to the tangent plane, and to the contour through $\bar{x}$, and points in the direction of fastest increase of f per unit of Euclidean distance, which is to say that $\nabla f(\bar{x}) \,.\, y$ is maximized, for all y such that $|y| = 1$, by choosing $y = \nabla f(\bar{x})/|\nabla f(\bar{x})|$.

It is worth noting that the gradient is not the direction of fastest increase if some other metric is chosen. If for example $|y| = \Sigma_j |y_j|$, then the best direction is obtained by finding that component of $\nabla f(\bar{x})$ having largest absolute value and setting the corresponding component of y at either $+1$ or -1 according to the sign of the component, and the remainder at zero. This is essentially the direction used in the simplex method for linear programming. Another norm is $|y| = \max_j |y_j|$, leading to the direction $y_j = \text{sgn}\,(\nabla f(\bar{x})_j)$, probably less useful.

7. *Quadratic approximations*

Ignoring the third-order terms of **21**, define the quadratic approximation F to f in the neighborhood of $\bar{x}$ by

$$F(x) = f(\bar{x}) + \nabla f(\bar{x})(x - \bar{x}) + \tfrac{1}{2}(x - \bar{x})\nabla^2 f(\bar{x})(x - \bar{x}), \qquad \mathbf{22}$$

the quadratic function agreeing with f through the second derivative terms. Since $F(x) - f(\bar{x}) - \nabla f(\bar{x})(x - \bar{x}) = \tfrac{1}{2}(x - \bar{x})\nabla^2 f(\bar{x})(x - \bar{x})$, the nonnegativity of one side entails that of the other, which establishes the connection between the characterizations **3** and **4** of convex functions in section 2, and is readily converted into a proof.

The expression for F can be simplified by 'completing the square': one can find p so that, making the change of variable $y = x - p$,

$$F(y) = q + ry^1 + \tfrac{1}{2}y^2 Q y^2, \qquad \mathbf{23}$$

where the vector r and the square matrix Q are of order summing to n, and y^1 and y^2, of the same orders, constitute a partition of the variables of y into two disjoint sets. In case $\nabla^2 f(\bar{x})$ is nonsingular there is no linear residue y^1, and $p = [\nabla^2 f(\bar{x})]^{-1} f(\bar{x})$. Thus to study a quadratic function, we may as well just look at

$$f(x) = -xQx. \qquad \mathbf{24}$$

This function is concave if and only if Q is positive semidefinite. Disregarding the fact that its maximum is known (at $x = 0$), the essential behavior of most processes for maximizing f can be studied with this case, as we do extensively in the sequel.

There are no transformations of the variables of a mathematical programming problem which leave all its important aspects invariant. For many purposes the most extensive change permitted is the 'scale change', the replacement of x_j by $k_j x_j$, where $k_j > 0$. This will not change the interpretation of the problem, reflecting only a change of the unit of the activity which

x_j measures. Unfortunately, even the scale change can seriously alter the behaviour of a simple algorithm.

Consider the maximization of $-x_1^2 - x_2^2$ under the constraints $0 \leqslant x_1, x_2 \leqslant 1$, using the maximal values of x_1, x_2 as point of departure. $\nabla f(1,1) = (-2, -2)$ points directly at the solution, the origin. Now make a scale change, letting $x_1 = 2\bar{x}_1$. Then $f(x_1, x_2) = -4\bar{x}_1^2 - x_2^2 = F(\bar{x}_1, x_2)$, say. The point of departure is $\bar{x}_1 = \frac{1}{2}$, $x_2 = 1$, and $\nabla F(\frac{1}{2}, 1) = (-4, -2)$, which no longer points at the origin but runs into the x_2-axis at $(0, \frac{3}{4})$; see Figure 3. The gradient may, of course, be appropriately transformed so as to point at the solution of the quadratic problem, but this is expensive. Such extensions will be discussed in section 9.

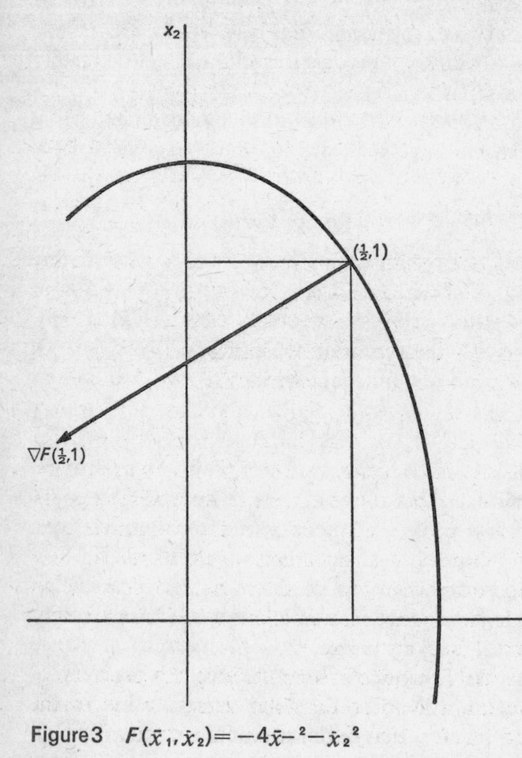

Figure 3 $F(\bar{x}_1, \bar{x}_2) = -4\bar{x}-{}^2 - \bar{x}_2{}^2$

Unconstrained Optimization Problems

8. *Unconstrained and one-dimensional problems*

Here we shall be concerned with procedures for maximizing f, assumed concave where convenient, under no constraints: S is the entirety of some linear manifold E^k. While problems really having this form are not common, there are two reasons for taking them up explicitly. On one hand, a number of techniques for the constrained problem rely entirely on procedures of this kind, merely making changes in f to handle the constraints (see 'Methods for Linear Constraints', p. 249). On the other hand, in the final stages of solving a problem with linear constraints generally the face of the constraint set on which the solution lies, or some close to it, has been determined, and the problem reduces to one of maximizing f on the manifold containing that face; its boundaries may be ignored.

The simplest problem of this kind is the maximization of f in one dimension – along a line. In many procedures this problem must be solved many times in the course of solving the main problem. While its efficient solution certainly depends on detailed knowledge of the particular function at hand, some useful remarks can be made.

Common practice is the use of a parabolic fit to f. Three data are needed. One certain to be at hand is the value $f(x)$ of the current best estimate, x, of the solution. If first and second derivatives along the line are available, Taylor's series through terms of second order would be practical. If only the first derivatives are available, then the directional derivative of f along the line, and the value of f at some other point on the line, are sufficient. Otherwise the value of f at two other points could be used. In any case a second-degree expression approximating f is obtained, and the point at which it obtains its maximum is used as the next best estimate – after comparison with $f(x)$ to find whether the approximation was good enough. There may be cases in which higher-order approximations would be more successful, but we feel them to be rare.

For quasiconcave functions of one variable the problem of finding a 'minimax' procedure for determining the extremum has been solved (Kiefer, 1957). The procedure, which depends rather delicately on the Fibonacci numbers, minimizes the amount of work required to find the maximum in the worst case. The following process is simpler, and nearly as good. Assume at a typical step that the maximum is known to lie in the interval $(a, a + h)$, and that the function values are known at $a + r^2h$ and $a + rh$, where $r = \frac{1}{2}(\sqrt{5} - 1)$. (Note that $r^2 + r = 1$ and $r^3 + 2r^2 = 1$.) The maximum then lies in $(a, a + rh)$ or $(a + r^2h, a + h)$ according to whether $f(a + r^2h)$ or $f(a + rh)$ is larger. Noting that $1 - r^2 = r$ we see that the length of the identified interval has been decreased by the factor r, and that f must be evaluated at only one new point (either $a + r^3h$ or $a + (1 - r^3)h$) for the next step.

9. *Ascent methods in general*

Most of the procedures for solving the unconstrained maximization problem are conveniently described as consisting in the recursion of the step **25**: For $k = 0, 1, 2, \ldots$, starting with arbitrary x^0,
given x^k, choose the *direction* Δx^k and the *step* length t^k. Let

$$x^{k+1} = x^k + t^k \, \Delta x^k. \qquad \mathbf{25}$$

In discussing a single step of the procedure we will drop the index ^{k} of iteration and indicate the step as

$$\bar{x} = x + t \, \Delta x. \qquad \mathbf{26}$$

There seem to be three general ways to choose Δx: essentially independently of the behavior of f in the neighborhood of x (such procedures are often called 'direct search'); in the direction of $\nabla f(x)$ (called 'gradient methods'); and in a direction dependent on $\nabla f(x)$ but intended to improve on that direction ('modified gradient' methods).

Some typical direct search proposals are: Choose Δx at random; choose Δx as at the previous iteration if that yields an increase in f, otherwise at random; choose Δx at a random acute angle with $\nabla f(x)$. Similar rules are used for the choice of t. Little of theoretical interest can be said about these proposals, although some experimental work has been done (Hooke and Jeeves, 1961; Wheeling, 1960). We shall not discuss them further.

A gradient method chooses $\Delta x = \nabla f(x)$. This is sometimes called the 'method of steepest ascent', and will be discussed in the next two sections. It has been the most widely used and studied procedure, since choosing Δx as the direction of fastest increase of f certainly seems like a good thing. The gradient direction is best, however, only locally; as that direction is pursued, it becomes steadily worse (see Figure 3 in section 7). The difficulty is an essential one, and is perhaps what distinguishes extremization problems with nonlinear objectives most sharply from linear problems.

The third class of methods modifies the gradient in order to find a better direction of motion: $\Delta x = H(x)\, \nabla f(x)$, where $H(x)$ is a suitable matrix. The quadratic approximation **22** of section 7 has its maximum at the point $\bar{x}$ obtained by letting $H(x) = [\,\nabla^2 f(x)]^{-1}$ and $t = 1$. If f is quadratic, then the problem is solved in one step, which argues for the power of this method, although the calculation is tedious. When done at each step of the gradient method for a non-quadratic problem it constitutes 'Newton's method'. Possible variations are a search along the line $x + t\Delta x$ for a better value of t than $\frac{1}{2}$, or possibly using the same value of $H(x)$ in several iterations despite the change of x. When the second derivatives are not available, the interesting approach of Davidon (1959) should be of value: successive values of $\nabla f(x^k)$ are used to build an estimate of $\nabla^2 f(x)$ and thus of the above $H(x)$. None of the modified gradient methods has yet been widely exploited.

10. *The 'optimal gradient' method*

A gradient method chooses $\Delta x = \nabla f(x)$ in equation **26**; the various gradient procedures differ in their choice of t. The most thoroughly studied has been the so-called 'optimal' gradient method for which $t = t_m$, the value of t achieving

$$\max \{f(x + t\, \nabla f(x)) : t \geqslant 0\}, \qquad \textbf{27}$$

so that

$$\bar{x} = x + t_m\, \nabla f(x). \qquad \textbf{28}$$

Note should be taken of an interesting property of the method: successive gradients $\nabla f(x)$, $\nabla f(\bar{x})$ are orthogonal, because the derivative with respect to t of the maximand of **27** at t_m is precisely $\nabla f(\bar{x}) \cdot \nabla f(x)$, which is required to vanish. Speaking more loosely, we shall also call the procedure 'optimal gradient' if t_m yields only a local maximum along the line $x + t\,\nabla f(x)$, $t \geqslant 0$, which is also a global maximum for $0 \leqslant t \leqslant t_m$. With either notion, Curry (1944) has proved for continuously differentiable functions the convergence of the process in the sense of this theorem:

Theorem: If $\nabla f(x)$ is continuous, then any limit of a sequence of points given by the optimal gradient procedure is a stationary point.

The essential idea of the proof is that if the limit point were not a stationary point then higher values of f would exist in the neighborhood of the limit point and would be obtained by the process.

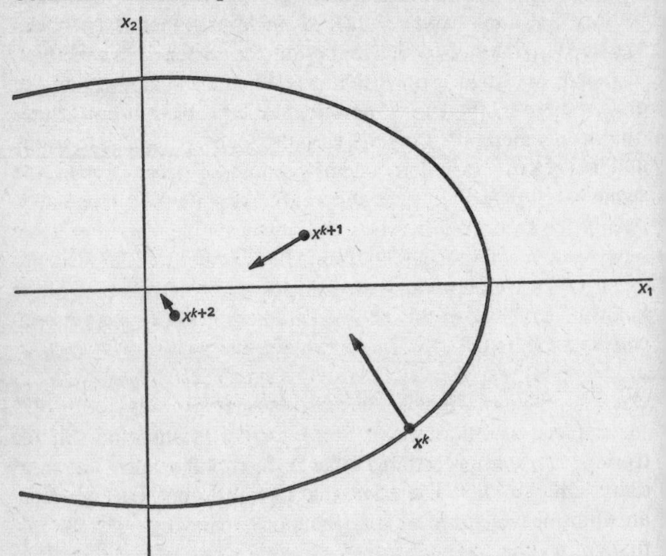

Figure 4 Optimal gradient method

Figure 4 illustrates how the optimal gradient process minimizes $f(x) = x_1^2 + 4x_2^2$. Since successive gradients are orthogonal, that at x^{k+2} is always parallel to that at x^k, and since the gradient is linear in x the points x^k, x^{k+2} lie on one of two lines passing through the origin, convergence to which is geometric.

In more than two dimensions the parallelism of successive gradients does not hold, but at least for positive definite quadratic forms convergence is still geometric, the point x^k drawing closer to the minimum by a factor of at least $(M - m)/(M + m)$, on the average, at each iteration (see Kantorovich, 1952, p. 109), where M and m are the largest and smallest eigenvalues of the quadratic form.

11. *Improvements on the optimal gradient method*

It might seem that the choice $t = t_m$ of **28** for the point along the gradient direction specified by the optimal gradient method, which is by definition best for the one-dimensional problem **27**, would be best for the entire procedure, but that is not the case – the best tactics do not always constitute the best strategy. Indeed, for the two-variable quadratic problem (Figure 4) the best choice of t would be that for which $\bar{x}$ falls on one of the principal axes of the quadratic form, for then the next step would solve the problem. Except in the circular case $\bar{x}$ always overshoots one axis and falls short of the other, which suggests use of 'subrelaxation', suggested by M. R. Hestenes (Stein, 1951): a constant ρ, $0 < \rho < 1$, is chosen, and the step is

$$\bar{x} = x + \rho t_m \nabla f(x). \qquad \mathbf{29}$$

Some early experimental work reported by Stein (1951) shows considerable improvement in speed of convergence with suitable ρ. The problem studied was that of minimizing a convex quadratic expression in six variables. The choice $\rho = 0{\cdot}9$ was most effective.

Another modification of the optimal gradient method, introduced by Forsythe and Motzkin (1952), seems to be even more successful. We call it a 'diagonal step' procedure. Its instructions are the same as those of the optimal gradient method, except that from time to time instead of using the gradient direction we let

$$\Delta x^k = x^k - x^{k-2}; \qquad \mathbf{30}$$

the direction is that of the third side of the triangle whose vertices are x^{k-2}, x^{k-1} and x^k.

The procedure is naturally suggested by the two-variable quadratic problem (Figure 4), which it solves immediately;

the minimum lies on the diagonal ray. For the general quadratic problem, x^{k+1} yields the minimum of the function on the two-dimensional manifold containing the three previous points. (There does not seem, however, to be any direct generalization of the procedure which would solve a higher-dimensional quadratic problem in a finite number of steps.) Experiments have been made on the same problems solved with subrelaxation, with even better results (Forsythe and Motzkin, 1952). The best performance came from taking a diagonal step after about seven optimal-gradient steps, resulting in an increase in speed of convergence of a factor of about 20 over the optimal gradient method.

The recent 'method of parallel tangents' (Shah, Buehler and Kempthorne, 1964) seems to hold even more promise for the convex nonlinear problem: using little more data than does the optimal gradient method, it finds the minimum of a quadratic function in a finite number of steps. (Of course, direct solution of the quadratic problem by inversion of the quadratic form is also finite, but has no direct analog in the non-quadratic case, as do these procedures.)

In the method of parallel tangents the instructions for Δx^k are of two kinds, but the new point is always chosen 'optimally': to achieve $\max_t \{f(x^k + t\Delta x^k) : t \geqslant 0\}$.

Start: Choose x^0, let $\Delta x^0 = \nabla f(x^0)$, and choose the next point 'optimally' – call it x^2.

Cycle k: Given x^{2k-2} and x^{2k}:

Let $\Delta x^{2k} = \nabla f(x^{2k})$, choose x^{2k+1} 'optimally';

Let $\Delta x^{2k+1} = x^{2k+1} - x^{2k-2}$, choose x^{2k+2} 'optimally'.

(Note that the rules do not generate a point x^1.)

The proof (Shah, Buehler and Kempthorne, 1964) of termination of the procedure in the quadratic case is long and not very intuitively appealing. Its essence is that the vectors $y^k = x^{2k} - x^{2k-2}$ form, for $k = 1, \ldots, n$, a conjugate system for the quadratic form Q – that is, that $y^j Q y^k = 0$ for $j \neq k$. Since a conjugate system can have only n members different from zero, the procedure terminates after n cycles entailing $2n - 1$ one-dimensional optimizations.

Incorporation of Constraints

12. *A differential method*

For theoretical purposes the method of steepest ascent can be stated in terms of the differential equation

$$\frac{dx}{dt} = \nabla f(x), \tag{31}$$

describing continuous motion of the point $x(t)$. For computational purposes we would of course rely on a discrete version of **31** such as **25**, but as we have seen better methods than those suggested directly by **31** are available, particularly since the trajectory $x(t)$ is of itself of no interest but only the hoped-for-solution $\lim_{t\to\infty} x(t)$. The same comment applies to the equations **32** below formulated for the constrained maximization problem of section 5, the maximization of $f(x)$ subject to $g_i(x) \leqslant 0$, $i = 1, \ldots, m$; but the equations have an interesting theory. They are

$$\frac{dx}{dt} = \nabla f(x) - \sum_i h_i(x)\, \nabla g_i(x), \tag{32}$$

where

$$h_i(x) = \begin{cases} 0 & \text{if} \quad g_i(x) \leqslant 0, \\ K & \text{if} \quad g_i(x) > 0, \end{cases} \tag{33}$$

and K is a suitably large positive number.

Since $\nabla g_i(x)$ is normal to the boundary surface $g_i(x) = 0$ and points away from S, the effect of the term $h_i(x)\, \nabla g_1(x)$ is to kick x back into the constraint set if it tends to leave it under the influence of $\nabla f(x)$.

Suppose that ∇f and all ∇g_i are continuous and that **32** have a solution $x(t)$ for which $\lim_{t\to\infty} x(t) = x^*$ and $x^* \in S$ – that is, $g_i(x^*) \leqslant 0$ for all i. If $g_i(x^*) < 0$, then clearly $h_i(x(t)) = 0$ for sufficiently large t; but if $g_i(x^*) = 0$ then $h_i(x(t))$ may oscillate forever between 0 and K as $x(t)$ is kicked about in a shrinking, but never stationary, orbit. Let us anyway assume that h_i has an average value for large t:

$$u_i^* = \lim_{T\to\infty} \frac{1}{T} \int_T^{2T} h_i(x(t))\,dt \quad \text{for all } i.$$

Integrating both sides of **32**, we have

$$x(2T) - x(T) = \int_T^{2T} \nabla f(x(t))dt - \sum_i \int_T^{2T} h_i(x(t))\, \nabla g_i(x(t))dt.$$

Dividing both sides by T and letting $T \to \infty$, we have finally

$$0 = \nabla f(x^*) - \sum_i u_i^* \,\nabla g_i(x^*), \qquad \mathbf{34}$$

or

$$\begin{cases} \nabla f(x^*) = \sum_i u_i^* \,\nabla g_i(x^*), \quad u_i^* \geqslant 0, \\ u^* \quad = 0 \ \text{ if } \ g_i(x^*) < 0. \end{cases} \qquad \mathbf{35}$$

Geometrically stated, **35** asserts that $\nabla f(x^*)$ belongs to the cone spanned by the normals $\nabla g_i(x^*)$ to those constraint surfaces $g_i(x) = 0$ on which x^* lies. Equation **34** states that the values u_i^* are exactly those needed to restrain x to the constraint set. Equations **35** are the Lagrange multiplier conditions of Kuhn and Tucker (1951), and it is easy to verify that they constitute conditions for a saddle point of the Lagrangian

$$L(x, u) = f(x) - \sum_i u_i g_i(x). \qquad \mathbf{36}$$

13. *A sequential method*

The idea of using the constraints of the problem to modify its objective is an old one in mathematical programming. The modification of the objective is to be such that, for points satisfying the constraints, the original objective is changed only little, while for points outside, or tending to be outside, the constraint set, the modified objective has very low values. A solution of the modified problem, without regard to the constraints, should then be nearly a solution of the original problem.

Let G be a concave nonincreasing function of one variable y, defined for $y < 0$ (and possibly for $y \geqslant 0$ also). The modified objective is

$$F(x, r) = f(x) + r \sum_i G(g_i(x)), \qquad \mathbf{37}$$

which is concave in x for $r \geqslant 0$ if f is concave and the g_i are convex (see section 2).

Section 12 dealt with a special case in which we set $r = K$

and chose $G(y) = \min\{0, -y\}$. The differential equations 32 were equivalent to $dx/dt = \nabla_x F(x, K)$, with the nondifferentiability of G at 0 requiring a special description of the process. Most of the constraint-incorporation procedures studied have used a differentiable G, defined only for $y < 0$, which keeps the point x in the interior of the constraint set. We will deal here only with these, for which

$$\lim_{y \to 0-} G(y) = -\infty. \qquad 38$$

The two most popular cases are $G(y) = y^{-1}$ and $G(y) = \log(-y)$.

We assume that the constraint set has an interior, and that for each $r > 0$ a point x_r maximizing $F(x, r)$ can be found. The procedure then consists just in maximizing $F(x, r)$ for sufficiently small r, and accepting x_r as the solution; obviously, all the types of solutions discussed in section 4 may appear. The question of how small r must be has to be settled *post hoc.*

If it is supposed that $\max_x F(x, r)$ has been solved precisely, the value of x_r as a solution can be estimated. Since x_r is a stationary point for the unconstrained modified problem,

$$0 = \nabla_x F(x_r, r) = \nabla f(x_r) + r \sum_i G'(g_i(x_r)) \nabla g_i(x_r),$$

or

$$\nabla f(x_r) = \sum_i u_i \nabla g_i(x_r), \qquad 39$$

letting $u_i = -rG'(g_i(x_r)) \geqslant 0$. Then for any y in S (concavity of f and convexity of g_i are assumed)

$$f(y) \leqslant f(x_r) + \nabla f(x_r)(y - x_r) = f(x_r) + \sum_i u_i \nabla g_i(x_r)(y - x_r)$$

$$\leqslant f(x_r) + \sum_i u_i[g_i(y) - g_i(x_r)] \leqslant f(x_r) - \sum_i u_i g_i(x_r),$$

so that the maximum value of f lies between $f(x_r)$ and $f(x_r) - \sum_i u_i g_i(x_r)$. (In terms of the duality theory for nonlinear programming (Wolfe, 1961), u constitutes a feasible solution of the dual, whose objective – the upper bound found above – dominates that of the primal.)

In using the procedure one determines from this bound whether x_r is acceptable and, if not, reduces r and solves again.

For certain methods of maximizing F the knowledge of the reasonably good starting point x_r will be useful. Since it appears that maximizing $F(x, r)$ becomes more difficult the smaller r is, because then the change of its gradient near the boundary of the constraint set becomes more abrupt, it is beneficial for that problem to use as large a value as possible; while the smaller r is, the closer x_r will be to the desired solution. The precise manner, then, of choosing successive r is of considerable importance, but has not yet been adequately formalized.

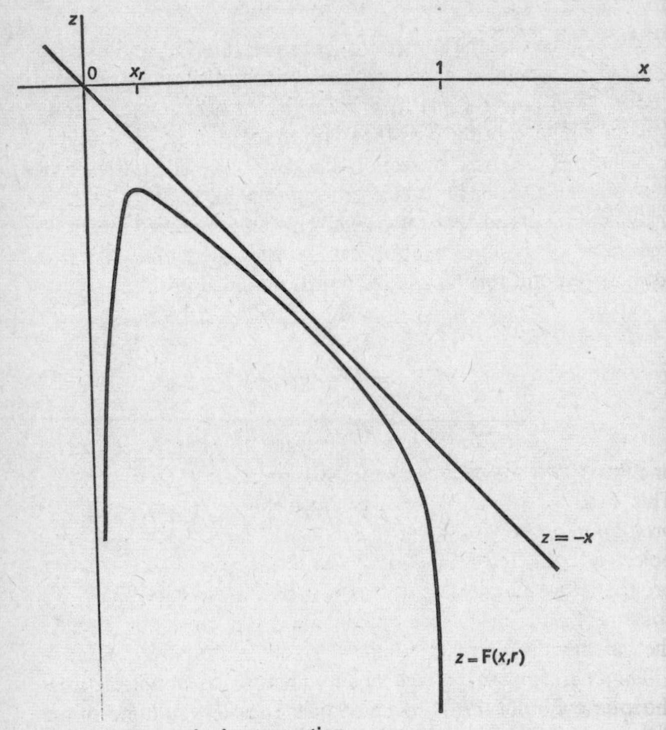

Figure 5 Constraint incorporation

Figure 5 illustrates maximizing $f(x) = -x$ under the constraints $0 \leqslant x \leqslant 1$, that is, with $g_1(x) = -x \leqslant 0$ and $g_2(x) =$

$x - 1 \leqslant 0$. Then $F(x,r) = -x + r[G(-x) + G(x-1)]$, which is maximized when $-G'(-x) + G'(x-1) = 1/r$. If r is small then $-G'(-x)$ is large, since $G' \leqslant 0$, and thus x_r is small. Ignoring the second term, approximately $G'(-x_r) = -1/r$, and $x_r \to 0$ as $r \to 0$, so the problem is solved.

The version using $G(y) = y^{-1}$ has been studied most extensively computationally, by Fiacco and McCormick (1963). They report good results on a variety of nonlinear problems, while stressing the importance of continued experimentation with the manner of choosing successive values of r, the relative efficiencies of various methods of finding x_r, the degree to which rough approximations to x_r can be used, and other matters.

Frisch (1955) and Parisot (1961) have studied the method using $G(y) = \log(-y)$, primarily for the case of linear objectives. Frisch calls it the 'logarithmic potential method', Parisot 'logarithmic programming'. It is not clear that there would be much difference in the operation of procedures having, as do these mentioned, quite similar penalty functions G.

Methods for Linear Constraints

14. *Problems with linear constraints*

When all the constraints g_i are linear, we write them as

$$g_i(x) = \sum_j a_{ij} x_j - b_i \leqslant 0, \qquad i = 1, \ldots, m, \qquad 40$$

or in matrix form $Ax \leqslant b$, where A is m by n and b is m by 1. This case is of interest because much of the theory of linear programming applies. It may be possible to approximate the objective by a linear function and use the very efficient linear programming algorithms; but even if that cannot be done, the constraint set is much easier to handle computationally than in the general case.

The constraint set $S = \bigcap_i \{x : g_i(x) \leqslant 0\}$ is a polyhedron, the intersection of the m indicated halfspaces, each of which has the boundary $\{x : g_i(x) = 0\}$. Let the dimension of S be d ($d = n$ is not necessary, although it is of course typical). For any family of boundaries, their intersection with S is called a *facet* of S. The $d - 1$ dimensional facets are called the *faces* of S; the one-dimensional facets are its edges, and the zero-dimensional facets its vertices. S is itself a facet, the intersection of S with the empty family of boundaries.

Solving a problem with linear constraints may be thought of as having two parts: the determination of the smallest facet on which the solution lies – the *solution facet* – and finding the solution in that facet. Unless the solution facet is a vertex, the solution lies in its relative interior. Restricting the objective, then, to the smallest linear manifold containing that facet gives us an essentially unconstrained problem for which a certain local maximum is to be found; and if the objective is such that any local maximum is global, the solution of the problem on the manifold is the solution desired. Thus once the solution facet has been identified the constraints play little role, and all the possibilities for efficient solution of the unconstrained problem (discussed in 'Unconstrained Optimization Problems' p. 239) can be used. Actually the solution facet is of course not known in advance, but is better and better identified as the algorithm proceeds.

There are two main ways of handling the constraints computationally. In the first procedure discussed below, the reduced gradient method, the devices of the simplex method for linear programming are used; in the second, a direction of motion is confined to a chosen facet by being projected upon it.

In this chapter we shall assume that $m \geqslant n$, so that there are enough boundaries to intersect in at least one vertex of S, and further that any p by n submatrix of A above has rank p for $p \leqslant n$. It is easy then to show that S is nondegenerate: exactly n of its boundaries meet in any vertex. (This assumption is stronger than is needed for the sequel, but is the most convenient one. The devices needed in case of degeneracy are like those used with the simplex method, eg. Wolfe, 1963).

15. *The reduced gradient method*

In the reduced gradient method the constraints are handled as in the simplex method for linear programming. This may not be immediately clear, owing to the statement of our problem; the connexion with the simplex method will be related later.

Under our hypotheses, any n-rowed submatrix of A has rank n

Let A be partitioned into such a submatrix C and the remaining rows, D, with the constants b_i of **40** similarly partitioned into c and d. The constraints **40** may then be written $Cx \leqslant c$, $Dx \leqslant d$, or, using the slack variables y (n in number) and z ($m - n$ in number), as

$$\begin{cases} Cx + y = c \\ Dx + z = d \end{cases} \qquad y, z \geqslant 0. \qquad \mathbf{41}$$

We will use the variables y, instead of x, as a set of co-ordinates for the space; this is possible since C is nonsingular. The variables z will be treated as dependent. The conditions $(y, z) \geqslant 0$ characterize the points belonging to S, and the face to which the point belongs is determined by those components of (y, z) which vanish. The vertices are characterized by the vanishing of n components; by hypothesis, no more than n can vanish. The independent variables y will always be chosen so that those which vanish, if any, are included among them, and the remainder are positive. This choice is essential, for we are then assured that any motion Δy such that $\Delta y_j \geqslant 0$ if $y_i = 0$, if not made too large, will not cause any of the dependent variables z to become negative. We thus have a convenient means of moving from the given point in any direction within S. If the motion terminates in the vanishing of an independent variable, then a new direction is chosen. If it terminates in the vanishing of a dependent variable, then that variable is declared independent, one of the non-vanishing independent variables is declared dependent, and a new direction is chosen.

The simplex method follows the above description, altering only one independent (nonbasic) variable at a time. When a dependent (basic) variable vanishes, the above interchange is made; it is the 'pivot operation' of the simplex method. With a linear objective, if any increase in a variable is profitable, it may be increased until some variable vanishes, and thus starting at a vertex where n variables vanish we move to another where n vanish, throughout the problem. In the nonlinear case the motion may stop short, so that such basic solutions are not of importance; and we may enjoy the freedom of letting several independent variables change simultaneously. Note that from **41** we have, taking y independent,

$$\left.\begin{aligned} \Delta x &= -C^{-1}\Delta y, \\ \Delta z &= DC^{-1}\Delta y, \text{ and} \\ \nabla_y f(x) &= -\nabla f(x)C^{-1} \text{ (the 'reduced gradient').} \end{aligned}\right\} \quad \mathbf{42}$$

One step of the procedure is as follows:

(1) Let $\Delta y_j = \begin{cases} [\nabla_y f(x)]_j \text{ if } [\nabla_y f(x)]_j > 0 \text{ or } y_j > 0, \\ \text{zero otherwise.} \end{cases}$

(2) If $\Delta y = 0$, the problem is solved. Otherwise, find Δx and Δz from **42**.

(3) Let $\theta_1, \theta_2, \theta_3$ be respectively the smallest values of θ achieving

$$\begin{aligned} &\max \{\theta : y + \theta\Delta y \geqslant 0\}, \\ &\max \{\theta : z + \theta\Delta z \geqslant 0\}, \\ &\max \{f(x + \theta\Delta x)\}, \end{aligned}$$

all with $\theta \geqslant 0$. Let $t = \min \{\theta_1, \theta_2, \theta_3\}$, and

$$\begin{aligned} \bar{x} &= x + t\Delta x, \\ \bar{y} &= y + t\Delta y, \\ \bar{z} &= z + t\Delta z. \end{aligned}$$

(4) If $t < \theta_2$, return to (1). Otherwise choose a vanishing z to be independent and the largest y_j to be dependent, and perform the corresponding pivot operation on the data of **42**. Return to (1).

Step (1) actually admits of considerable variation. The formula given is the analog of the ordinary gradient method for unconstrained problems, but any direction in which f increases would serve. Some of the acceleration devices for the unconstrained problem can be used here; the diagonal step procedure (section 11), occasionally replacing $\nabla_y f(x)$ by $y^k - y^{k-2}$ in step (1), has proved successful in practice (Huard and Wolfe, in press), and seems to make the method quite efficient.

16. *The gradient projection method*

Given the feasible point x, the smallest facet containing x is the intersection of the – say, p – boundaries for which $g_i(x) = 0$. Of course $p \leqslant n$; we may have $p = 0$, in which case x is interior to S. Let M be the p by n matrix consisting of the corresponding p rows of A. Since M has rank p, MM^T is nonsingular. Let $P = I - M^T(MM^T)^{-1}M$ (in case $p = 0$, set $P = I$). Then fo any vector y, the vector Py is the orthogonal projection of y onto the intersection of the p linear manifolds $\{z : g_i(z) = 0\}$, that is, onto the linear manifold $\{z : Mz = 0\}$. (That follows from the observations that for any y, $MPy = 0$, and that if $Mz = 0$ then $Pz = z$; P is linear, maps the whole space into the manifold, and maps the manifold onto itself, so is a projection; it is orthogonal because $P(I - P) = 0$.)

A single step of the gradient projection method is as follows:

(1) Let $\Delta x = P\,\nabla f(x)$.

(2) If $\Delta x \neq 0$, find $t^* = \max\,\{t : x + t\Delta x \varepsilon S\}$ and t_m, maximizing $f(x + t\,\Delta x)$ for $0 \leqslant t \leqslant t^*$. Let $\bar{x} = x + t_m\,\Delta x$. If some additional $g_i(\bar{x})$ vanishes, modify M. Return to (1).

(3) If $\Delta x = 0$, find $u_{ir} = [(MM^T)^{-1}M\,\nabla f(x)]_r$ for $r = 1, \ldots, p$.

(a) If all $u_{ir} \geqslant 0$, stop; x is the solution.

(b) Otherwise delete from M the row corresponding to the most negative of u_{ir}, and return to (1).

In step (1), the direction Δx is chosen as the projection of the gradient on a certain facet containing x. In step (2), if the projection does not vanish, $\bar{x}$ is taken as the best point in S in the direction Δx. If $t_m < t^*$, there is no change in M; but if $t_m = t^*$ then another boundary plane has been reached, and another constraint must be added to those of M; the dimension of the working facet has been thus decreased by one.

In step (3) the projection vanishes. Define the m-vector u by letting $u_i = u_{ir}$ for $i = i_r$ and $u_i = 0$ otherwise. Then $P\,\nabla f(x) = 0$ implies $\nabla f(x) = M^T u = \sum_i u_i\,\nabla g_i(x)$, the Lagrangian condition. Thus the problem is solved if $u \geqslant 0$. If however, some $u_i < 0$ then it can be shown that f can be increased by moving from x in the direction of decreasing $g_i(x)$; so the corresponding temporary constraint $g_i(x) = 0$ is deleted, yielding a facet of dimension one greater in which a direction of increase can be found.

Figure 6 sketches the course of the method for the problem of finding the closest point in the polyhedron to P. Points x^1 and x^2 are obtained from a step (2) with $t_m = t^*$, and x^3 with $t_m < t^*$. From x^3 a step (3) is required; x^4, the solution, results.

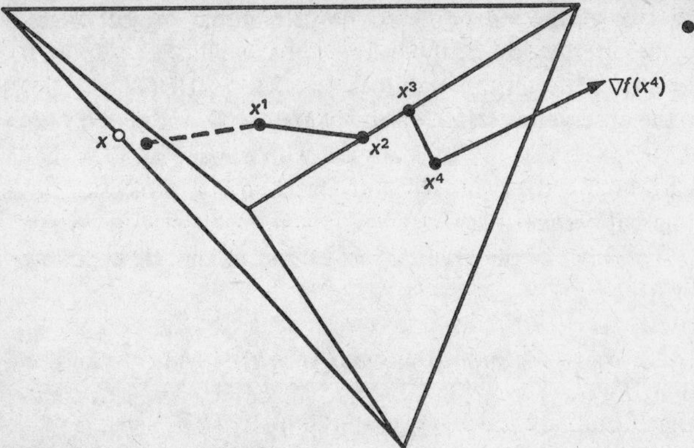

Figure 6 Gradient projection

(Precisely in the case of the minimum-distance problem the method terminates in a solution, since projection preserves the ordering of Euclidean distances.)

Linear Approximation Methods

17. *On columnar methods*

In this chapter we shall confine our attention to methods for the general nonlinear problem which make good use of our ability to solve linear programming problems efficiently. Besides the procedures of 'Incorporation of Constraints' (p. 245) we thus exclude other recent methods of interest, such as those of Graves (in press) and Wilson (1964), which use other devices and whose theory is less closely related to the main themes of these chapters.

Both types of procedure investigated here make use of linear approximations to the problem, but in different ways. The first of these, the decomposition method presented in the next section, requires some preliminaries about the 'columnar methods' we shall now develop.

Let $x^1, x^2, \ldots, x^T$ be a collection of n-vectors. Any point of the convex hull of this collection may be written

$$x = \sum_{t=1}^{T} s^t x^t, \quad \text{where} \sum_t s^t = 1 \quad \text{and} \quad s^t \geqslant 0 \text{ (all } t). \qquad 43$$

Given any function h of x, the linearization of h on the *grid* $x^1, \ldots, x^T$ is attained through the approximation

$$h(x) \cong \sum_t s^t h(x^t). \qquad 44$$

Any problem becomes a linear problem in the variables s^t if x and each function of x are replaced by their representations above. Our general problem becomes:

Maximize $\sum_t s^t f(x^t)$ subject to

$$s^t \geqslant 0, \sum_t s^t = 1, \text{ and } \sum_t s^t g_i(x^t) \leqslant 0 \text{ (all } i). \qquad 45$$

Let $\bar{s}^1, \ldots, \bar{s}^T$ be the solution of this problem. Then

$$\bar{x} = \sum_t \bar{s}^t x^t \qquad 46$$

is offered as the approximate solution of the original problem. How well this approximation fits will depend of course on the fineness of the grid $x^1, \ldots$, but also in an important way on the nature of the functions f and g_1 or upon the manner in which the linear programming problem **45**, **46** has been handled.

If the functions g_i are convex, then

$$g_i(\bar{x}) = g_i\left(\sum_t \bar{s}^t x^t\right) \leqslant \sum_t \bar{s}^t g_i(x^t) \leqslant 0,$$

so that $\bar{x}$ belongs to S. If f is concave, then

$$\sum_t \bar{s}^t f(x^t) \leqslant f(\bar{x}),$$

so that the maximal objective of the linear problem bounds the value for $\bar{x}$ on one side, although an estimate of the true maximum is not available. When f is strictly concave and all g_i are convex the solution of the linear programming problem has this important property: There is a solution of the linear programming problem such that those grid points having positive weights s^t are the vertices of a convex polyhedron containing no other grid point. In the absence of this property, $\bar{x}$ would not be a valid interpolation among the grid points – if, for example, x^1, x^2, x^3 lay along a line and x^1 and x^3 had positive weights while x^2 had none. That property entails that in the one-dimensional case only two grid points will have positive weights, and they must be adjacent. If f is further strictly concave, then the linear problem has only solutions of this kind.

When the suitable convexity properties do not obtain it is nevertheless possible to assure that $\bar{x}$ is a valid interpolation by insisting on the simplex property for solutions of the linear problem, which can be done by a suitable restriction on the simplex method when solving it. The development of such a scheme has been accomplished under the title of 'separable programming', discussed in detail elsewhere in this volume. The restrictions on the solution make it impossible, of course, to be sure of obtaining a global solution of, even, the linear problem; but the local solutions obtainable are of great value, and separable programming has become perhaps the most powerful of the currently used techniques for nonlinear problems whose functions have certain fairly simple forms.

18. *The decomposition procedure*

In the case of several dimensions, any grid of reasonable fineness covering a large region will contain a tremendous number of points, rendering the linear problem **45** intractable. Actually only a few of the data of that problem would ever be used in finding the solution, however, which itself will involve only $m + 1$ columns of data from **45**. The decomposition algorithm for linear programming problems of special structure (Dantzig and Wolfe, 1961) uses a particular technique for generating only those data which are needed in the calculation which is readily extended to the nonlinear problem, conceived as being represented by a linear problem of the form **45** constructed from an arbitrarily fine grid.

Let a grid be given, and the problem **45** solved, yielding as well as the solution $\bar{s}^1, \ldots$ the dual solution $u_0, u_1, \ldots, u_m$ (where the equation $\sum_t s^t = 1$ is numbered '0' and the remainder $1, \ldots, m$). We pose the question: Of all points x^{T+1} that might be added to the grid as a further refinement, which would the simplex method first choose as contributing the most to the solution of the thus extended linear problem?

The column added to the data of **45** will have the objective coefficient $f(x^{T+1})$ and the remaining coefficients $(1, g_1(x^{T+1}), \ldots)$. Its 'reduced cost' will be then $f(x^{T+1}) - u_0 - \sum_i u_i g_i(x^{T+1})$; the simplex method chooses the column with maximal reduced cost. The new point is thus given by the solution of the problem

$$\text{Maximize } f(x) - u_0 - \sum_i u_i g_i(x), \ x \text{ unconstrained.} \qquad \mathbf{47}$$

If it should happen that the maximum value is not positive, then nothing more can be gained by adding points, and the problem is solved. Otherwise we let x^{T+1} be this solution x, and construct a column $T+1$ to add to the linear problem, and solve it again.

The repeated application of this procedure constitutes the decomposition procedure for the nonlinear problem. It is demonstrably successful only when f is concave and the g_i convex. It can be shown that an estimate of the maximal objective is always at hand; $f(x)$ is of course a lower bound, and the gap between $f(x)$ and the maximum is never greater than the value found in **47**. Roubault (1964) has shown the convergence of the process to a solution of the nonlinear problem as $T \to \infty$.

Solving and resolving the linear problem is routine. In this procedure, the burden of work has been shifted to the problem **47**, which may itself be difficult; of course, a precise maximization is not needed – only a point x giving a positive value to the objective of **47**. A popular special case is that of separability of the functions f and g_i:

$$f(x) = \sum_j f_j(x_j), \qquad g_i(x) = \sum_j g_{ij}(x_j),$$

a not uncommon case. Then the maximization of **47** falls into n separate maximizations, each of

$$f_j(x_j) - u - \sum_i g_{ij}(x_j),$$

giving problems which are likely to have nearly trivial solutions.

19. *Constraint approximation methods*

The columnar procedures were based on the idea of representing S as the convex hull of a set of points; the constraint approximation methods may be thought of as representing it as the intersection of a family of halfspaces. To make best use of our ability to solve linear programming problems, we wish to linearize the objective function in the same way; so we shall suppose, as we have shown possible in section 5 without loss of generality, that f is linear.

Let a set of points $x^1, \ldots, x^T$ be given, and suppose (only for the sake of definiteness in the following discussion) that associated with each is some index i_t between 1 and m. For each t we shall replace the constraint $g_{i_t}(x) \leqslant 0$ by its first-order Taylor's series approximation. The nonlinear problem then becomes:
Maximize $f(x)$ subject to

$$g_{i_t}(x^t) + \nabla g_{i_t}(x^t)(x - x^t) \leqslant 0, \qquad t = 1, \ldots, T. \qquad \mathbf{48}$$

We shall suppose all the g_i are convex. Then every point of S satisfies the constraints of **48**, because the left-hand side of **48** is never greater than $g_{i_t}(x)$. The constraint set of **48** may be much larger than S, and indeed the procedure will be efficient just insofar as **48** yields a good solution of the original problem without requiring a great number T.

As with the decomposition procedure, only a few – namely, n – of the relations of **48** are needed to define its solution; and, when a solution has been found, we can find that inequality to add which will do the most good. Thus let the solution of **48** be x^{T+1}, and define i_{T+1} as that index i for which $g_i(x^{T+1})$ is maximal. The *cutting-plane* procedure of Kelley then consists simply in the adjunction of the new constraint $T+1$ to the system and the resolving of the new linear programming problem. This step is repeated until a solution is obtained which satisfies the constraints $g_i(x) \leqslant 0$ to a sufficient degree. (If any solution should happen to satisfy all the constraints, then the process would stop. The solution would also be the solution of the original problem, since it would maximize the objective over a constraint set at least as large as the original.) Two steps of the procedure are illustrated in Figure 7, using a linear objective which increases toward the right.

The convergence of the procedure is not difficult to prove. The necessary starting condition is that an initial set of points can be chosen so that the objective of the linear problem is bounded above.

Note that the point x^{T+1} always lies outside the constraint set, and, indeed, does not satisfy the constraint constructed from it; substituting $x = x^{T+1}$ in the $T + 1$ constraint of **48** gives $g_{iT+1}(x^{T+1})$, which is positive by construction. The added con-

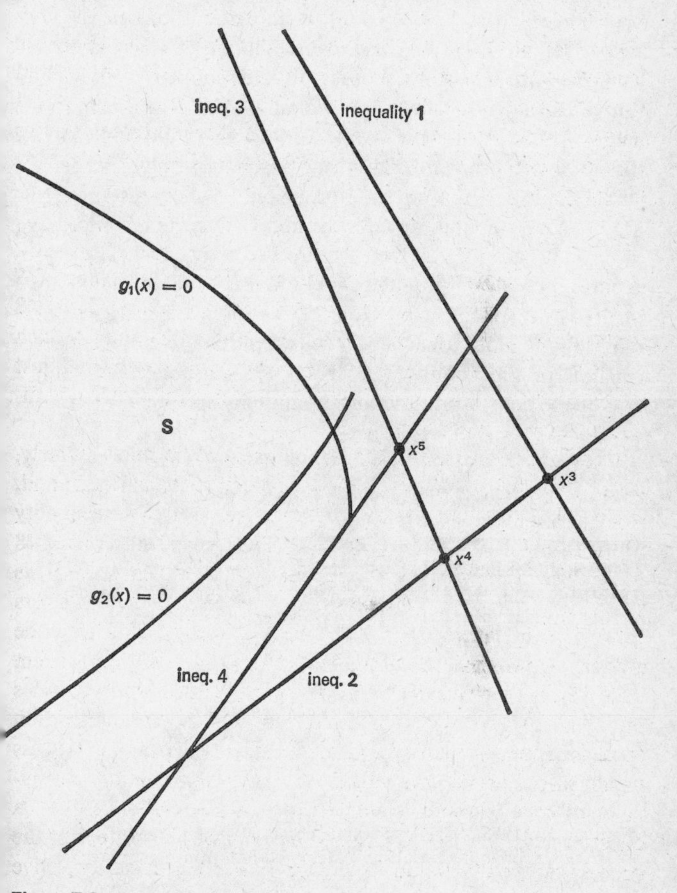

Figure 7 Cutting-plane method

straint thus 'cuts' away the former solution x^{T+1} of the linear problem, producing an improved approximation to the original constraint set in its neighborhood. The fact that only one constraint is not satisfied when the new linear problem is to be solved makes use of the dual simplex method very cheap in the procedure.

Mention should be made of two other procedures for this problem which make use of linear approximations to the constraint set. Rosen (1961) and Zoutendijk (1960) have proposed using first-order Taylor's series approximations to the constraint functions to define a linear programming problem which, in the neighborhood of the current best solution, would approximate the nonlinear problem. The linear problem is solved under the added constraint that its solution be only a short distance from the old solution, and the new solution is corrected, if necessary, to bring it back into S; then the process is repeated. These procedures, like the cutting-plane method (and unlike the columnar methods in general, it appears) have their work increased only modestly if additional linear constraints are added to the original problem, and seem therefore advisable when only a few of the constraint functions are nonlinear.

References

CURRY, H. B. (1944), 'The method of steepest descent for non-linear minimization problems', *Q. Appl. Math.*, vol. 2, pp. 258–60.

DANTZIG, G. B., and WOLFE, P. (1961), 'The decomposition algorithm for linear programs', *Econometrica*, vol. 5, pp. 767–78.

DAVIDON, W. C. (1959), *Variable Metric Method for Minimization*, Argonne Nat. Lab. Report No. ANL-5990.

FENCHEL, W. (1953), *Convex Cones, Sets, and Functions*, Princeton University, Department of Mathematics.

FIACCO, A. V., and MCCORMICK, G. P. (1963), *Programming under Nonlinear Constraints by Unconstrained Minimization: A Primal-Dual Method*, Research Analysis Corporation, RAC-TP-96.

FORSYTHE, A. I., and MOTZKIN, G. E. (1952), *I.B.M. Experiments with Accelerated Gradient Methods for Linear Equations*, Nat. Bureau Standards Report, no. 1643.

FRISCH, R. (1955), 'The logarithmic potential method for solving linear programming problems', Memorandum from the University Institute of Economics, Oslo.

GRAVES, G. (in press), *A Nonlinear Programming Procedure*, The RAND Corporation.

HOOKE, R., and JEEVES, T. A. (1961), '"Direct search" solution of numerical and statistical problems', *J. Assoc. Computing Machinery*, vol. 8, pp. 212–20.

HUARD, P., and WOLFE P. (in press), *The Reduced Gradient Method.*

KANTOROVICH, L. V. (1952), *Functional Analysis and Applied*

Mathematics, translated by C. D. Benster, Department of Mathematics, University of California.

KELLEY, J. E., Jr, (1960), 'The cutting-plane method for solving convex programs', *J. Soc. Indust. Appl. Maths*, vol. 8, pp. 703–12.

KIEFER, J. (1957), 'Optimum sequential search and approximation methods under minimum regularity assumptions', *J. Soc. Indust. Appl. Maths*, vol. 5, pp. 105–36.

KUHN, H. W., and TUCKER, A. W. (1951), 'Nonlinear programming', *Proc. Second Berkeley Symp. on Maths, Stat. and Prob.*, University of California Press, pp. 481–92.

PARISOT, G. R. (1961), 'Résolution numérique approchée du problème de programmation linéaire par application de la programmation logarithmique', Thesis, Université de Lille.

ROSEN, J. B. (1960), 'The gradient projection method for nonlinear programming: Part I, linear constraints', *J. Soc. Indust. Appl. Maths*, vol. 8, pp. 181–217.

ROSEN, J. B. (1961), 'The gradient projection method for nonlinear programming, Part I, nonlinear constraints', *J. Soc. Indust. Appl. Maths*, vol. 9, pp. 514–32.

ROUBAULT, M. C. (1964), 'De la dualité des programmes mathématiques convexes', Abstract, *1964 International Symposium on Maths Programming*, London.

SHAH, B. V., BUEHLER, R. J., and KEMPTHORNE O. (1964), 'Some algorithms for minimizing a function of several variales', *J. Soc. Indust. Appl. Maths*, vol. 12, pp. 74–92.

STEIN, M. L. (1951), *Gradient Methods in the Solution of Systems of Linear Equations*, Nat. Bureau Standards Report, no. 52–7.

WHEELING, R. (1960), *Studies in Optimization, Progress in the Optimization of Nonlinear Functions*, Socony Mobil Oil Co., Research Department, Paulsboro, New Jersey.

WILSON, R. (1964), 'Duality methods for mathematical programming', Thesis, Harvard University.

WOLFE, P. (1961), 'A duality theorem for non-linear programming', *Q. Appl. Maths*, vol. 19, pp. 239–44.

WOLFE, P. (1963), 'A technique for resolving degeneracy in linear programming', *J. Soc. Indust. Appl. Maths*, vol. 11, pp. 205–11.

ZOUTENDIJK, G. (1960), *Methods of Feasible Directions*, Elsevier, Amsterdam.

14 A. Charnes and W. W. Cooper

Deterministic Equivalents for Optimizing and Satisficing under Chance Constraints

A. Charnes and W. W. Cooper, 'Deterministic equivalents for optimizing and satisficing under chance constraints', *Operations Research*, vol. 11, 1963, pp. 18–39.

Chance constrained programming admits random data variations and permits constraint violations up to specified probability limits. Different kinds of decision rules and optimizing objectives may be used so that, under certain conditions, a programming problem (not necessarily linear) can be achieved that is deterministic – in that all random elements have been eliminated. Existence of such 'deterministic equivalents' in the form of specified convex programming problems is here established for a general class of linear decision rules under the following three classes of objectives: (1) maximum expected value ('E model'); (2) minimum variance ('V model'); and (3) maximum probability ('P model'). Various explanations and interpretations of these results are supplied along with other aspects of chance constrained programming. For example, the 'P model' is interpreted so that H. A. Simon's suggestions for 'satisficing' can be studied relative to more traditional optimizing objectives associated with 'E' and 'V model' variants.[1]

The topic of chance constrained programming is perhaps best introduced by first exhibiting an ordinary linear programming problem in its general form as

$$\max c'x$$

with $$Ax \leqslant b, \qquad \mathbf{1}$$

where A is an $m \times n$ matrix of constants and c, b are corresponding constant vectors. The problem is then to choose a set of values[2] for the variables of vector x so that (a) they satisfy

1. 'Satisficing' is a term used by Simon – e.g., (see Simon, 1957 and 1959) to contrast some of his characterizations with the optimizing propensities that are often assumed in economics, operations research, and elsewhere. (The term was, however, also accorded dictionary status – 'v.t., to satisfy, *Scot.* and *Dial. Eng.*' – until the most recent issue of *Webster's International* dropped it as obsolete.)

2. Non-negativity requirements, if any, are assumed to be incorporated in the constraints of **1**.

all constraints and (b) render $c'x$ a maximum in accordance with the given criterion elements $c' = (c_1, c_2, \ldots, c_n)$ and the stated objective of maximization.

An Introduction to Chance Constrained Programming

A chance constrained formulation would replace **1** with a problem of the following kind:

$$\text{optimize } f(c,x)$$

subject to $$P(Ax \leqslant b) \geqslant a, \qquad \mathbf{2}$$

wherein 'P' means 'probability'. Here A, b, c are not necessarily constant but have, in general, some or all of their elements as random variables. The vector a contains a prescribed set of constants that are probability measures of the extent to which constraint violations are admitted. Thus, an element $0 \leqslant a_i \leqslant 1$ is associated with a constraint $\sum_{j=1}^{j=n} a_{ij} x_j \leqslant b_i$ to give

$$P\left(\sum_{j=1}^{j=n} a_{ij} x_j \leqslant b_i\right) \geqslant a_i, \qquad \mathbf{3}$$

a double inequality which is interpreted to mean that the ith constraint may be violated, but at most $\beta_i = 1 - a_i$ proportion of the time.

There is, as might be expected, a rather wider range of reasonable choices to be considered for **2**, compared with **1**, when deciding upon an objective. Since this paper is concerned with exploring a variety of possible objectives, we shall, for the moment, leave this aspect of **2** formulated only in a very general way. Then, as we shall see, a choice of objectives may carry with it certain implied kinds of constraint consequences at least when judged from the standpoint of equivalent problems.

We shall assume that a choice of values for the decision variables, x, will not disturb the densities associated with the random variables in A, b, c. Then we may formulate the general problem in terms of choosing a suitable decision rule

$$x = \varphi(A,b,c) \qquad \mathbf{4}$$

with the function, φ, to be chosen from a prescribed class of functions and applied in a manner that guarantees that the x values, as generated, will (a) satisfy the chance constraints of **2** and (b) optimize $f(c,x)$ in **2** by reference to the class of rules from which the φ of **4** is to be chosen.

In some cases we may wish to delimit the class from which φ is to be chosen so that the resulting x values are assigned numerical values. When this can be done we shall say that the resulting φ^* provides a set of 'certainty equivalent relations'. In other cases we may permit a choice of φ in which the x vectors are also random. Thus, we may wish to study the distributions of the x components as functions of the original random variables or we may wish to specialize further and study only the distribution of functional values $z^* = f(c,x)$. In the latter class of cases we may refer to φ^* as a stochastic relation because of the close relation that it bears to the kind of problem posed by G. Tintner under the name 'stochastic linear programming'.[3] In either the stochastic or certainty equivalent case it may also be desirable – and possible – to develop a deterministic equivalent for an originally stated chance constrained problem at least as a way station for convenient analyses, characterizations, or solutions. These deterministic equivalents, when attainable, may also be valuable in their own right, for purposes of policy review, theoretical analyses, etc.

Some Chance Constrained Concepts and Terminology

These ideas probably need some further elaboration because, as we shall see, the choice of rules leads to different ways of interpreting the constraints and, also, the way the rule is applied requires some consideration. To effect a reasonably concrete explanation we shall have recourse to an abbreviated (and simplified) version of the problem that gave rise to the first chance constrained programming formulation and application (see Charnes, Cooper and Symonds, 1958).

Hereafter we shall specialize the constraints of **2** by assuming that the matrix, A, is constant. We shall also specialize the rule **4** to members of the class

3. See, e.g., Tintner (1960 and 1955).

$$x = Db \qquad 5$$

where D is an $n \times m$ matrix whose elements are to be determined by reference to 2.

The originally considered problem of scheduling heating oil to an uncertain – but known statistically, via its weather dependency – demand (Charnes, Cooper and Symonds, 1958) may be abbreviated by leaving out some of its parts. It would then appear as

$$\min E \sum_{j=1}^{j=n} (c_j R_j + k_j \bar{I}_j),$$

$$\text{subject to} \quad P\left(I_0 + \sum_{j=1}^{j=l} R_j \geqslant \sum_{j=1}^{j=l} S_j\right) \geqslant a_l \quad (l = 1, \ldots, n), \qquad 6$$

$$\text{and} \qquad P(R_j \geqslant 0) \geqslant 1 \qquad (j = 1, \ldots, n).$$

Here the symbol 'E' refers to the expected value operator so the objective is an expected value optimization to be considered over a horizon of $j = 1, \ldots, n$ periods. The c_j and k_j are, respectively, the refinery costs and inventory carrying charges that are allowed to be random. I_0 is the initial inventory and S_j ,a random variable, represents the sales for which the refinery rates, R_j, are to be determined so that – as the constraints indicate – all sales are to be met at prescribed a_l levels and no negative R_j are to be permitted.

We shall refer to the a_l as 'risk levels' even though a more precise usage might restrict this to the $\beta_l = 1 - a_l$ values. Then we shall distinguish these from the $b_t = \sum_{j=1}^{j=l} S_j$ part of the double inequality. The latter will be called 'quality levels'. In the present case these levels cover the full magnitude of all sales. Obviously, however, this need not always be the case. For instance, the original random variables may be scaled, or otherwise adjusted, by suitable constants so that $b_t = \kappa_l \hat{b}_l \pm t_l \sigma_l$ may refer to a certain fraction of the original sales variables, say, to which a known amount is to be added or subtracted. The problem 6 would then alter in no theoretically essential way, but in period l the random variables would be altered so that, say, 95 per cent of all sales demands plus or minus a safety margin, would be met, say, 98 per cent of the time. Notice, also, that by merely adjusting and iterating any constraint we can secure still further controls, if desired. For example, with the indicated safety margin we may wish to satisfy 95 per cent of the sales demand that emerges, say, 98 per cent of the time, 96 per cent of the sales 93 per cent of the time, and so on.

In dealing with **6** it is desired to arrange matters so that the actual refinery rates will be known exactly by the start of the period to which they apply. In short, a certainty equivalent relation – or set of relations – is wanted. For this we specialize the rule of **5** to

$$D = (\Gamma, \delta) \qquad \mathbf{7}$$

where Γ is lower triangular. More precisely, we arrange the matrix of **7** so that it gives relations of the form

$$\sum_{j=1}^{j=l} R_j = \sum_{j=1}^{j=l} \gamma_{lj} S_j + \delta_l, \qquad (l = 1, 2, \ldots, n) \qquad \mathbf{8}$$

with each γ_{lj}, δ_l a scalar and $\gamma_l \equiv 0$ for each l.

Here matters can be arranged so that the γ_{lj} and δ_l values can first be determined via a suitably arranged deterministic equivalent for **8**. After these values have been secured the refinery rates are set as follows. At the start of period 1 set $R_1 = \delta_1$. Then at the start of period 2, after S_1 has been observed, choose $R_1 + R_2 = \gamma_{21} S_1 + \delta_2$. At the start of period 3, when S_1 and S_2 have both been observed, choose $R_1 + R_2 + R_3 = \gamma_{31} S_1 + \gamma_{32} S_2 + \delta_3$. And so on.

Observe that *different* γ_{tj} weights apply to the *same* observations. Given the values of the γ_{jl}, δ_l coefficients – e.g. as determined via a suitable deterministic equivalent – then **8** supplies a set of *relations* that can be applied to generate the needed (for certain) *numbers*, as required.[4] Of course, to complete the application in this case the observations on the preceding S_j values must be obtained. All parts of the relation, as needed for specifying the R_j values, will then be known with certainty. The

4. The issue of existence here deals with *relations* and is also distinguished in other ways from the standard theorems on the existence of certainty equivalent *numbers*. Cf., e.g. Holt *et al.* (1960, ch. 6).

resulting R_j values will, of course, then also be known with the certainty that this problem requires.

The indicated certainty equivalents may supply all that is required for the conduct of operations. It does not of itself, however, necessarily meet all of the needs of management. For instance, a management may want to evaluate different alternatives before committing itself to a given course of action or set of policies. But the above certainty equivalent cannot be completed until after the S_j values have been observed.

When a suitably arranged deterministic equivalent is available, however, it can be used as a basis for the wanted initial evaluations. Suppose, for example, that the deterministic equivalent assumes the form of an ordinary linear programming problem or, more generally, that it can be shown to form a convex programming problem. The sharp duality relations of either ordinary linear programming or the more general convex programming variety[5] are then available to supply such 'evaluators'. Then, in advance of the data needed for certainty equivalent operations, it is possible to study the over-all consequences of variations in risk levels, quality levels, etc., along with other forms of constraint alteration and data testing.

In this paper we propose to examine important classes of chance constrained problems and to obtain deterministic equivalents that are then shown to be convex programming problems.[6] It is to be emphasized, however, that optimization under risk immediately raises very important questions concerning a choice of rational objectives. Questions can arise, for example, concerning the reasonableness of an expected value optimization. Without attempting to resolve these issues, we should note that the evaluators secured for one objective are not necessarily correctly applied to the same problem under an altered objective. Hence, we shall examine three different classes of objectives that have some fairly evident claims to importance and relate these to the formulation given in **2**, above. This includes (i) an expected value optimization, (ii) a minimum variance (or mean-square

5. See Charnes, Cooper and Kortanek (1962 and 1963) for these results on the general convex programming problem.

6. We shall not go forward here, however, to a characterization of the resulting duality relations. This is reserved for treatment in a subsequent paper.

error) objective, and (iii) a maximum probability model. We shall refer to these as (i) an 'E model', (ii) a 'V model', and (iii) a 'P model'. It is of possible interest to distinguish between the first two cases and the third by reference to what H. A. Simon (Cf., e.g., 1957 and 1959) calls 'satisficing' – as contrasted with 'optimizing' – objectives. That is, we shall interpret our case **3** as a version of a satisficing model but we shall reduce it to the form **2**, for purpose of comparison, even though Simon originally proposed this objective as an alternative that might help to circumvent the inadequacies of optimizing objectives and probability calculations for characterizing important aspects of human behavior. The interpretation that we shall accord to satisficing under our P model may not adequately handle all aspects of what Professor Simon has in mind. Nevertheless it will bring out important aspects of constraint alterations that occur in our satisficing model (a) that have an interesting bearing on the objectives in our first two optimizing models, and (b) that do not appear in the other cases. Finally, to further emphasize these objectives we shall continue to employ the rule **5** and, in particular, we shall assume that the criterion elements and the indicated objective are all adequately stated in the functionals. In particular we shall avoid introducing any elements of the c vector into the constraints via **5** but we shall, rather, let these develop naturally as the deterministic equivalents are formed.

Choice of Decision Rules and Constraint Interpretations

The developments of the following sections, which are fairly abstract and general, are intended to cover any possible choice of a D matrix, as in **5**. That is, we shall be concerned with the general class of all possible rules of this kind. It may not be immediately evident that any restriction on the D choices is closely associated with a way of interpreting the constraints and so we had perhaps now best illustrate what is involved here.

Consider, for instance, the special D for **8** when applied to **6**. This leads to a constraint interpretation of the following kind:

$$P\{I_l + R_l \geqslant \underset{I_0}{\underline{S_l}} + \sum_{j=1}^{l-1} R_j - \sum_{j=1}^{l-1} S_j \equiv I_l\} \geqslant a_l . \quad (l=1, \ldots, n) \qquad \mathbf{9}$$

That is, given the preceding observations and decisions – on the right of the stroke – and, in advance of S_l, we are to determine a rate that when added to beginning inventory, I_l, for period l, will be adequate to cover the unknown S_l with at least the specified probability a_l.

Evidently other D choices will be associated with different ways of interpreting the constraints. Some consistency requirement with the A matrix of **2** is therefore evidently imposed. But consistency with A is not, alone, decisive.[7] The problem context, the kinds of policy conditions to be honored, and even the preferences (perhaps not wholly rational) of the responsible decision makers must all be considered.

We can secure some further illumination by reference to work done by A. Ben-Israel (1962 and, with Charnes, 1961) on what we shall call the zero order decision rule. Here we can consider *any* A and assume that a 'decision maker' wants to know all of his x_j programme values in advance. This is supposedly true, for example, in some aspects of country development programming (by three to five year plans) as well in other kinds of budgetary planning practice. Hence the case is of some interest and importance and, fortunately, it leads to a solution by reference to straightforward fractile calculations. As might be expected it also collapses the 'certainty' and 'deterministic equivalent' categories into a single case.

The development can be briefly summarized as follows. Choose, say, the rule **7** and set $\Gamma = 0$, the null matrix. This gives $Ix = I\delta$ and so we may as well work with the x values directly. Thus, let $\psi(b_1, \ldots, b_m)$ be the multivariate density for b. Then under suitable assumptions,[8] we can obtain

$$\psi_i(b_i) = \int_{-\infty}^{\infty} \cdots \int_{-\infty}^{\infty} \psi(b_1, , \ldots b_m) \prod_{j=1,\, j \neq i}^{j=m} db_j \qquad \mathbf{10}$$

7. There is also the issue of choosing a rule (linear or not) that will be optimal for the problem of interest. See Charnes and Cooper (1959) for a discussion of how chance constrained programming may be 'turned around' to bear on this problem.

8. See Ben-Israel (1962, and Ben-Israel and Charnes, 1961) for precise characterizations and developments.

as the marginal density for the variate b_i. But then

$$\int_{-\infty}^{\tilde{b}_i} \psi_i(b_i)db_i + \int_{\tilde{b}_i}^{\infty} \psi_i(b_i)db_i = 1 \qquad 11$$

for any fractile $\tilde{b}_i$. Choosing $\tilde{b}_i = \tilde{b}_i(1 - a_i)$ to satisfy

$$\int_{-\infty}^{b_i} \psi_i(b_i)\, db_i = 1 - a_i,$$

we employ Ben-Israel's theorem and observe[9]

$$P(\sum_{j=1}^{j=n} a_{ij} x_j \leqslant b_i) \geqslant a_i \Leftrightarrow \sum_{j=1}^{j=n} a_{ij} x_j \leqslant \tilde{b}_i(1 - a_i).$$

The expression on the right is an ordinary linear programming constraint. Assuming that it forms part of a solvable system[10] of $i = 1, \cdots, m$ constraints derived in a precisely analogous manner we can carry the ordinary aspects of linear programming duality into play as follows. Let the initially stated objective for **2** in this class of cases assume the form optimize $f(c,x) = \max E\, c'x$ with c' a vector of random variables and

$$Ec'x = \mu_c' x \qquad 13$$
$$\mu_c' = (\mu_{c1}, \ldots, \mu_{cn})$$

so that μ_c' is a vector whose elements are the means of c'. The dual to the problem

$$\max \mu_c' x,$$

subject to $\quad Ax \leqslant \tilde{b}(1 - a), \qquad$ **14**

is $\quad \min w' \tilde{b}(1 - a),$

subject to $\quad w'A = \mu'c. \qquad$ **15**

But we know the density $\varphi(c_1, \ldots, c_n)$ for the variates in c. Hence we can determine a vector of stipulations β, via $P(c \leqslant \mu_c) \geqslant \beta$, and thus we can replace the constraints of **15** by

9. The symbol '$\Leftrightarrow$' means 'if and only if'.

10. See A. Ben-Israel (1962) for a precise elaboration of the abbreviated development that is given here.

$$P(w'A \geqslant c') \geqslant \beta'. \qquad 16$$

By proceeding on lines like these Ben-Israel utilizes these zero order rules and obtains two ordinary linear programming problems – viz,

$$\begin{array}{llll} & \max c'(\beta)x, & & \min w'\tilde{b}(1-a), \\ \text{subject to} & Ax \leqslant \tilde{b}(1-a), & \text{subject to} & w'A \geqslant \tilde{c}'(\beta), \\ & x \geqslant 0, & & w' \geqslant 0, \end{array} \qquad 17$$

which gives the x and w solutions relative to the fractiles of the originally stipulated densities.

It would evidently be possible to extend this discussion further in order to examine other kinds of interesting and important classes of cases. We do not propose to do this.[11] Instead we want to examine all possible rules of the form D and, for important classes of objectives and statistical distributions, we want to be able to characterize situations in which a deterministic equivalent will be achieved – irrespective of the D choice – which yields a convex programming problem. This is done in the immediately following sections.

'E model'

Our expected value model is now presented in the form

$$\begin{array}{ll} & \max Ec'x, \\ \text{subject to} & P(Ax \leqslant b) \geqslant a, \\ & x = Db, \end{array} \qquad 18$$

where the same meanings as before apply to these symbols. A is, as we previously observed, assumed to be a matrix of known constants. The vectors b and c, however, may have any or all of their elements as random variables. For clarity and convenience in the presentation we shall also assume that b and c are uncorrelated, but this condition need not apply to their respective components. For instance, all parts of the analysis are applicable even when $Ec_j c_k \neq Ec_j Ec_k$, $Eb_i b_r \neq Eb_i Eb_r$, etc.[12]

11. See Appendix (p. 281) for further illustrations.

12. As will become clear subsequently, any correlation of c_j and b_i variates will still result in a linear objective function for the deterministic equivalent.

In order to achieve a deterministic equivalent for **18** we first substitute from $x = Db$ into the functional and write

$$E(c' Db) = (Ec)'D(Eb) \qquad \textbf{19a}$$

so that the functional elements are now all deterministic. Then we define the vectors

$$\mu_c' \equiv (Ec)'; \qquad \mu_b' \equiv (Eb)', \qquad \textbf{19b}$$

so that **19a** becomes $\mu_c'D\mu_b$ with the variables to be determined being impounded in D alone. The functional is now entirely deterministic.

We now effect a similar substitution in the constraints and achieve

$$\min - \mu_c' D\mu_b,$$

subject to $$P(ADb \leqslant b) \geqslant \alpha. \qquad \textbf{20}$$

The constraints still involve the vector b of random variables and hence this problem is not deterministic.

To facilitate the subsequent developments we introduce

$$\hat{b} \equiv b - \mu_b,$$
$$a_i' \equiv (a_{i1}, \cdots, a_{in}), \qquad \textbf{21}$$

so that a_i' is the ith row of A.

Henceforth we shall also assume (*a*) that the frequency distributions for the variates $(a_i'D\hat{b} - \hat{b}_i)$ are symmetric and (*b*) that the distributions associated with these variates are completely specified by their first two moments.[13] This does not necessarily mean that the distributions for the variates b and c must have these same properties and, of course, we have the transformations $a_i'D$ and other devices available for securing suitable approximations in other kinds of cases.[14]

For concreteness we assume that the variate $(a_i' Db - b_i)$ is

13. These conditions can be relaxed when the α_i are chosen to be suitably large. Analogous constructions are also possible when one parameter frequency functions are used, etc. Also, when D is specialized in various ways numerous other cases can be accommodated, but we do not propose to pursue these possibilities further at this time. See for example Ben-Israel (1962) and Ben-Israel and Charnes (1961).

14. Cf., for example Ben-Israel and Charnes (1961) on utilizing mixtures of normal distributions to approximate other (non-normal) distributions.

normally distributed because the normal distribution (*a*) is important in its own right and (*b*) has the properties indicated by our preceding assumption. Then we can write

$$\begin{aligned} P(a_i'Db - b_i \leqslant 0) &= P(b_i - a_i'\,Db \geqslant 0) \\ &= P(\hat{b}_i - a_i'D\hat{b} \geqslant -\,\mu_{bi} + a_i'D\mu_b). \end{aligned} \qquad 22$$

Assuming[15] $E[\hat{b}_i - a_i'D\hat{b}]^2 > 0$ we can divide both sides of the last expression by the positive square root and obtain

$$\begin{aligned} &P(\hat{b}_i - a_i'D\hat{b} \geqslant -\,\mu_{bi} + a_i'D\mu_b) \\ &\quad = P\left(\frac{\hat{b}_i - a_i'\,D\hat{b}}{\sqrt{E[\hat{b}_i - a_i'\,D\hat{b}]^2}} \geqslant \frac{-\,\mu_{bi} + a_i'\,D\mu_b}{\sqrt{E[\hat{b}_i - a_i'\,D\hat{b}]^2}}\right). \end{aligned} \qquad 23$$

Next we define

$$z_i \equiv \frac{\hat{b}_i - a_i'\,D\hat{b}}{\sqrt{E[\hat{b}_i - a_i'D\hat{b}]^2}}, \qquad 24$$

so that z_i has zero mean and unit variance. Direct substitution in **23** and reference to **20** produces

$$P\left(z_i \geqslant \frac{-\,\mu_{bi} + a_i'\,D\mu b}{\sqrt{E[\hat{b}_i - a_i'\,D\hat{b}]^2}}\right) \geqslant \alpha_i \qquad 25$$

for the *i*th constraint. This may be equivalently rendered, of course, by

$$F_i\left(\frac{-\,\mu_{b_i} + a_i'\,D\mu_b}{\sqrt{E[\hat{b}_i - a_i'\,D\hat{b}]^2}}\right) \geqslant \alpha_i, \qquad 26$$

where F_i is the cumulant for the marginal density of z_i.

Since z_i is $N(0, 1)$ we may utilize its symmetry properties in order to achieve a deterministic equivalent which is a convex programming problem as follows. First we assume that all $\alpha_i > 0{\cdot}5$ – as will generally be the case for applications to managerial policy problems[16] – and then we solve **26** to achieve

$$\frac{-\,\mu_{b_i} + a_i'D\mu_b}{\sqrt{E[\hat{b}_i - a_i'D\hat{b}]^2}} \leqslant F_i^{-1}(\alpha_i) \equiv -\,K_{\alpha_i}, \qquad 27$$

with assurance that $K_{\alpha_i} > 0$, all *i*.

15. This assumption is made only to simplify the development.

16. See Charnes and Cooper (1962a) and Charnes, Cooper and Symonds (1958) for further discussion.

Observing that each K_{α_i} is a known, fixed number we are evidently now in a position to achieve a deterministic equivalent for our '*E*' model. The objective, however, is to obtain an equivalent that will be convex. For this reason – and also because it will aid in subsequent interpretations – we will first split each of the constraints, **27**, into an equivalent pair that are identified, respectively, with the risk and quality stipulations that we wish to examine. For the latter purposes we introduce new variables, v_i, and then rewrite **27** as

$$-\mu_{bi} + a_i' D\mu_b \leqslant -v_i \leqslant -K_{\alpha i}\sqrt{[E(\hat{b}_i - a_i' D\hat{b})^2]} \leqslant 0, \quad \textbf{28a}$$

$$\text{or} \quad \mu_{bi} - a'_{bi} D\mu_b \geqslant v_i \geqslant K_{\alpha_i}\sqrt{[E(\hat{b}_i - a_i' D\hat{b})^2]} \geqslant 0, \quad \textbf{28b}$$

wherein the inequalities on the right[17] are a consequence of the assumptions guaranteeing $K_{\alpha i} > 0$, all *i*.

We can utilize the expressions **28a** directly. By virtue of the non-negativity assigned to all expressions between inequality signs in **28b** we can replace each expression by its square without altering the sense of the inequality. Hence we have

$$-a_i' D\mu_b - v_i \geqslant -\mu_{bi}, \quad \textbf{28c}$$

$$-K^2_{\alpha i} E[\hat{b}_i - a_i' D\hat{b}]^2 + v_i^2 \geqslant 0, \quad \textbf{28d}$$

with $v_i \geqslant 0$, as a pair equivalent to **27** for each *i*.

We can now write

$$\min -\mu_c' D\mu_b,$$

subject to

$$\mu_i(D) - v_i \geqslant 0,$$

$$-K^2_{\alpha i}\sigma_i^2(D) + K^2_{\alpha i}\mu_i^2(D) + v_i^2 \geqslant 0, \qquad (i = 1, \ldots, m) \quad \textbf{29}$$

$$v_i \geqslant 0,$$

where[18]

$$\sigma_i^2(D) \equiv E(a_i' Db - b_i)^2, \quad \textbf{30}$$

$$\mu_i^2(D) \equiv (\mu_{bi} - a_i' D\mu_b)^2.$$

17. We have used weak, rather than strict, inequality in order to accommodate cases where some of the variances may be zero.

18. The deriviation is via

$$\sigma_i^2(D) = E[a_i' D\hat{b} - \hat{b}_i]^2 = E[(a_i' Db - b_i) - (a_i' D\mu_b - \mu_{bi})]^2.$$

This is a deterministic equivalent for **18** – achieved via the indicated assumptions – in that the random elements are either factored out via the 'E' operator for the risk terms or are replaced by the means, μ_b, for the quality terms.

We now observe that this is a convex programming problem in the variables D and v. To see that this is so refer to **29** and **30** and observe that the expressions $\mu_i(D) - v_i \geqslant 0$ are the algebraic corresponds of a finite collection of half spaces. Therefore, their common intersection is defined by a polyhedral convex set. Each of the expressions

$$- K^2_{\alpha i}\, \sigma_i^2(D) + K^2_{\alpha i}\, \mu_i^2(D) + v_i^2 \geqslant 0$$

corresponds to one nappe and the interior of an elliptic hyperboloid. This, too, corresponds to a convex set. Since the intersection of convex sets is also convex, the wanted result is at hand. This deterministic equivalent for **18** is a convex programming problem.

We should observe that none of the variables v_i appear in the functional. These variables may therefore be used freely in the constraints whenever an advantage is thereby gained for the functional values (see Charnes and Cooper, 1962a).

We can also observe that no components of μ_c are transferred to the constraints whereas all of the μ_b values appear both in the functional and in the constraints. This result – as will subsequently be seen – does not always occur but it nevertheless lends itself to an easily made interpretation. For then

$$\mu_c' D \mu_b = \sum_j \sum_k \mu_{c_j}\, d_{jk}\, \mu_{b_k} \qquad \textbf{31}$$

evidently exhibits the d_{jk} variables as weighted transformation ratios relative to the expected returns μ_{c_j} and the expected levels μ_{b_k}.

For concreteness suppose that μ_{cj} is stated in terms of dollars per unit of product output while (as in the Appendix, p. 281) μ_{bk} is a machine capacity in hours of input available. Then the associated d_{jk} must in be units of output per hour of input, a so-called 'production coefficient', which we have here designated by reference to the more general category of transformation ratios whose interpretation will depend on the data of any particular problem.

Because these d_{jk} values impound risk adjustment features, however, they are not precisely the same in all detail as the transformation ratios in the technical literature of economic analysis. Witness, for this, the constraints of **29** the terms of which are defined by **30** and observe that these also contain d_{jk} values so that if, say, a_{ij} is a processing time in hours per unit and b_k is, again, in hours of machine capacity, the resulting multiplication gives an $a_{ij}\, d_{jk}\, b_k$ in hours of deviation relative to some other machine b_i, for the random variables, as well as a further adjustment relative to the mean μ_{b_i}, which is known for certain. The result in either case provides a statement of $v_i \geqslant 0$ in hours and this variable relates the pure quality level constraint involving only $\mu_i(D)$ to the risk constraint which also involves $\mu_i(D)$ multiplied by the factor K_{α_i}.[19] The pure risk component, on the other hand, appears as $K^2_{\alpha i}\, \sigma_i^2(D)$ where, of course, this $\sigma_i^2(D)$ is a generalized variance whose values may be improved by suitable choices of D.

'V model'

We need not repeat the preceding developmental details or interpretations in the case of our V model. Here we only want to show that this kind of objective yields, again, a deterministic equivalent which is a convex programming problem. Moreover, as we shall see, this change in the objective of optimization yields only a change in the deterministic functional; it does not alter any of the constraints that were previously attained in the E model.

For our V model we write

$$\min E(c'x - c^{0\prime}x^0)^2,$$

subject to

$$P(Ax \leqslant b) \geqslant \alpha, \qquad \textbf{32}$$

$$x = Db,$$

19. See Table I in the Appendix (p. 284) for an interpretation of the $v_i \geqslant 0$ values associated with the non-negativity conditions on the variables x_j. Note also that the v_i do not appear in the functional and hence, as a kind of generalized 'slack,' they may be used to advantage in improving functional values or, in some cases, to simplify the solution procedures. (See Charnes and Cooper (1962a) for an illustration.)

where the objective is to minimize a generalized mean square error. I.e., taking all relations between the c_j into account, we wish to minimize this measure of their deviations about some given preferred value $z^0 = c^{0\prime}x^0$.[20]

Employing a development that is wholly analogous to the preceding one, we achieve the following deterministic equivalent to **32**:

$$\begin{aligned} &\min V(D), \\ \text{subject to} \quad &\mu_i(D) - v_i \geqslant 0, \\ -K_{\alpha_i}^2\, \sigma_i^2(D) + K_{\alpha_i}\, &\mu_i^2(D) + v_i^2 \geqslant 0, \\ &v_i \geqslant 0, \end{aligned} \qquad \mathbf{33}$$

where the definitions of **30** apply and

$$V(D) \equiv E(c'\,Db - c^{0\prime}\,x^0)^2. \qquad \mathbf{34}$$

This deterministic equivalent is again a convex programming problem. In fact, the constraint sets are exactly the same as for **30**. Only the functional forms differ. Hence, if any differences appear in the values secured for d_{jk}, v_i, via **29** and **33**, the alternation between an 'E' and a 'V' objective must be the source of the difference.

'*P model*'

Bearing the results in mind for these two 'orthodox' versions of optimization, we now turn to a version of the satisficing approach that has recently been suggested by H. A. Simon.[21] In this

20. When z^0 is a weighted combination of the μc_j values then this reduces to a generalized variance. For special applications of this kind of objective to the problem of portfolio selection see H. Markowitz (1959).

21. See Simon (1957, chs. 14 and 15). See also March and Simon (1958) and the extensions and applications to problems of budgetary management as given in Cooper and Savvus (1961), Stedry (1960) and Stedry and Charnes (1962).

approach the $c^{0\prime}x^0$ components are specified relative to some set of values – e.g., as generated from an aspiration level mechanism – which an organism[22] will regard as satisfactory whenever these levels are achieved. Of course, when confronting an environment subject to risk, the organism cannot be sure of achieving these levels when effecting its choice from what it believes are the available alternatives. On the other hand, if it does not achieve the indicated $c^{0\prime}x^0$ levels or, more precisely, if it believes that it cannot achieve them at a satisfactory level of probability then the organism will either (a) reorient its activities and 'search' for a more favorable environment or else (b) alter its aspirations and, possibly, the probability of achieving them.[23]

In order to place this kind of approach in a context that will permit easy comparisons with our preceding results, we now interpret this satisficing approach in the context of an optimizing model as follows:

$$\max \mathrm{P}[c'x \geqslant c^{0\prime}x^0],$$

subject to

$$\mathrm{P}[Ax \leqslant b] \geqslant a, \qquad 35$$

$$x = Db.$$

That is, we assume that the objective is to maximize the probability of achieving the specified $c^{0\prime}x^0$. Then, to bring into prominence the search features that are possible in such a model, we can single out certain constraints. For instance, we may prescribe

$$\mathrm{P}[c'x \leqslant c^{0\prime}x^0] \geqslant a_0. \qquad 36$$

when only one over-all aspiration level is involved. Alternatively, we can elaborate this to, say,

$$\mathrm{P}[c_j x_j \geqslant c_j^0 x_j^0] \geqslant a_{0,j}, \qquad 37$$

when we wish to particularize this model for selected elements of the total aspirations. The main point is, of course, that the selection of x values which maximize the functional need not

22. An individual or a business firm, in the present context, although Simon extends his analysis to other kinds of organisms in order to base his analyses more firmly on the validated results of a variety of psychological experiments.

23. It can, of course, do some of each.

satisfy all of these constraints. When this occurs, a contradiction is present and the search or revision mechanism specified by Dr Simon will then presumably come into play.[24]

We now propose to utilize the same rules and assumptions as before to reduce this to a deterministic equivalent. Employing algebraic reductions and analyses that are analogous to those already given, we then achieve

$$\max v_0/w_0,$$

subject to

$$\begin{aligned} \mu_c' D\mu_b - v_0 &\geqslant \mu_{c0}, \\ -V(D) + w_0^2 &\geqslant 0, \\ \mu_i(D) - v_i &\geqslant 0, \\ -K^2_{\alpha_i}\sigma_i^2(D) + K^2_{\alpha_i}\mu_i(D) - v_i^2 &\geqslant 0, \\ v_i &\geqslant 0. \end{aligned} \qquad \textbf{38}$$

In this case we observe that the last pair of constraints are the same as before. Two new sets of constraints are generated for the deterministic equivalent in this P model and, as can be seen, they each incorporate the functionals that were encountered in the deterministic E and V equivalents, respectively. Evidently these new constraints emanate from the reoriented objective for the P model interpretation of a satisficing approach.

The function and objective in **38** now assume a minimax-like character in the sense that maximization of v_0/w_0 represents a striving towards co-operatively maximizing v_0 while minimizing w_0. This functional is more complex than before in that v_0/w_0 need not be either convex or concave.[25] The constraint set of **38** remains convex just as before, however, and this provides immediate access to the fractional programming theorems in Charnes and Cooper (1962b). It must be borne in mind, of course, that one can represent an arbitrary connected convex set in finite dimensional space by an infinite system of linear inequalities (see Charnes, Cooper and Kortanek, 1962 and 1963). Therefore

24. This raises some issues, of course, with respect to 'rational' search procedures. Dr Simon does not extend his analyses in these directions, but a possible mode of deployment may be suggested by Charnes and Cooper, (1958) – when the model therein specified is properly oriented with respect to chance constraints.

25. For other extensions and applications, see Stedry and Charnes (1962).

the arguments of Charnes and Cooper (1962b) apply and it is an immediate consequence that a local maximum will also be a global maximum for linear fractional functionals. By virtue of these characterizations, however, plus the fact that $w_0 > 0$ we can replace the fractional programming problem **38** by *one* convex programming problem, as follows:

$$\max \bar{v}_0$$

subject to

$$\begin{aligned} \mu_c' \bar{D}\mu_b - \bar{v}_0 &\geqslant t\mu_c^0, \\ -\bar{V}(\bar{D}) + \bar{w}_0^2 &\geqslant 0, \\ \bar{\mu}_i(\bar{D}) - \bar{v}_i &\geqslant 0, \\ -K^2_{\alpha_i}\bar{\sigma}_i^2(\bar{D}) + K^2_{\alpha_i}\bar{\mu}_i^2(\bar{D}) + \bar{v}_i^2 &\geqslant 0, \\ \bar{w}_0 &= 1, \\ t, \bar{v}_i &\geqslant 0, \end{aligned} \qquad \mathbf{39}$$

wherein

$$\bar{D} = tD, \qquad \bar{v} = tv, \qquad \bar{w}_0 = tw_0, \qquad \mathbf{39a}$$

and

$$\begin{aligned} \bar{V}(\bar{D}) &= E(c'\bar{D}b - tc^{0\prime}x^0)^2, \\ \bar{\sigma}_i^2(\bar{D}) &= E(a_i'\bar{D}b - tb_i)^2, \\ \bar{\mu}_i^2(\bar{D}) &= (t\mu_{b_i} - a_i'\bar{D}\mu_b)^2. \end{aligned} \qquad \mathbf{39b}$$

Again a convex programming problem has been achieved and so the indicated reduction for the P model is also now completed.

We can now conclude on a qualifying note by observing that the d_{jk}, v_i values resulting from an E, V, or P orientation need not differ in any particular example. Until the area of chance constrained programming is carried further into the ideas of model types and equivalences, however, it can only be said that this must depend upon the data of any particular application. Finally, we observe that constraints like **36** and **37** can also be carried over into the E and V models.[26] This does not, however, affect the main point of our general comparisons on an alteration in constraints for the P model insofar as the analysis was restricted to the one general class of decision rules, D, and exactly the same general class of constraints as for the E and V models.

26. E.g., to handle risk limitations on funds, etc. Cf., e.g., B. Naslund and A. Whinston (1962) for an application to the area of dynamic portfolio planning.

Appendix

Simple examples may help to provide a better understanding of the more abstract developments in the text of this article. For this we borrow the following illustration of a machine loading model from Charnes and Cooper (1961):

$$\begin{aligned} & \max \$1x_1 + \$\tfrac{1}{2}x_2, \\ \text{subject to} \quad & 3x_1 + 2x_2 \leqslant 12, \\ & 5x_1 \leqslant 10, \\ & -x_1 \leqslant 0, \\ & -x_2 \leqslant 0, \end{aligned} \tag{40}$$

where the c_j values in the functional represent unit profits, the a_{ij}'s are processing times (per unit), and the b_i are machine capacities (in hours) exhibited on the right. The last pair of inequalities are the standard non-negativity requirements, written in this fashion in order to conform to the representation implied in the expressions **1**, **2** ff.[27] in the preceding text.

This example is a special version of

$$\begin{aligned} & \max \mu_{c1} x_1 + \mu_{c2} x_2, \\ \text{subject to} \quad & a_{11} x_1 + a_{12} x_2 \leqslant \mu_{b1}, \\ & a_{21} x_1 + a_{22} x_2 \leqslant \mu_{b2}, \\ & -x_1 \leqslant 0, \\ & -x_2 \leqslant 0, \end{aligned} \tag{41}$$

which will now be developed in a way that shows in a simple – but sufficiently general – manner (a) that the equivalence claimed for **29** is, in fact, correct and (b) that any deterministic linear programming problem can also be viewed in chance constrained terms.

To bring these points into prominence we assume that $\sigma_{b_1}^2 = \sigma_{b_2}^2 = \text{cov}(b_1, b_2) = 0$ so that the distributions are degenerate with their entire masses concentrated at the means μ_{b_1}, μ_{b_2}. We demand, naturally, that the constraints be fulfilled with probability one.[28] Then we utilize our decision rule, D, in the form

27. See, e.g., footnote 2 on p. 262.

28. Presumably these conditions can be relaxed, if desired, to study parametric variations, to handle nonsolvability, etc.

$$\begin{Vmatrix} x_1 \\ x_2 \end{Vmatrix} = \begin{Vmatrix} d_{11} & d_{12} & d_{13} & d_{14} \\ d_{21} & d_{22} & d_{23} & d_{24} \end{Vmatrix} \begin{Vmatrix} \mu_{b1} \\ \mu_{b2} \\ 0 \\ 0 \end{Vmatrix}, \qquad 42$$

which provides direct access to the first set of expressions in **29** – the only ones needed here by virtue of the assumed degeneracy of these distributions.

A direct application of **42** in **30** and **29** now gives

$$\begin{aligned} &\max \mu_{c_1} d_{11} \mu_{b_1} + \mu_{c_1} d_{12} \mu_{b_2} + \mu_{c_2} d_{21} \mu_{b_1} + \mu_{c_2} d_{22} \mu_{b_2}, \\ &\text{with } a_{11} d_{11} \mu_{b_1} + a_{11} d_{12} \mu_{b_2} + a_{12} d_{21} \mu_{b_1} + a_{12} d_{22} \mu_{b_2} + v_1 \leqslant \mu_{b_1}, \\ &a_{21} d_{11} \mu_{b_1} + a_{21} d_{12} \mu_{b_2} + a_{22} d_{21} \mu_{b_1} + a_{22} d_{22} \mu_{b_2} \quad + \quad v_2 \leqslant \mu_{b_8}, \\ &\quad - d_{11} \mu_{b_1} - d_{12} \mu_{b_2} \qquad\qquad\qquad\qquad + v_3 \leqslant 0, \qquad 43 \\ &\qquad\qquad\qquad - d_{21} \mu_{b1} - d_{22} \mu_{b2} + v_4 \leqslant 0, \\ &\qquad v_1, v_2, v_3, v_4 \geqslant 0 \end{aligned}$$

This is an ordinary linear programming problem in the variables d_{ij}, v_i and hence has a dual which can be written

$$\begin{aligned} &\min w_1 \mu_{b1} + w_2 \mu_{b2} \\ \text{with} \quad &w_1 a_{11} \mu_{b_1} + w_2 a_{21} \mu_{b_1} - w_3 \mu_{b_1} = \mu_{c_1} \mu_{b_1}, \\ &w_1 a_{11} \mu_{b_2} + w_2 a_{21} \mu_{b_2} - w_3 \mu_{b_2} = \mu_{c_1} \mu_{b_2}, \qquad 44 \\ &w_1 a_{12} \mu_{b_1} + w_2 a_{22} \mu_{b_1} - w_4 \mu_{b_1} = \mu_{c_2} \mu_{b_1}, \\ &w_1 a_{12} \mu_{b_2} + w_2 a_{22} \mu_{b_2} - w_4 \mu_{b_2} = \mu_{c_2} \mu_{b_2}, \\ &w_1, w_2, w_3, w_4 \geqslant 0. \end{aligned}$$

With $\mu_{b1}, \mu_{b2} \neq 0$, this reduces to

$$\begin{aligned} &\min w_1 \mu_{b1} + w_2 \mu_{b_2}, \\ \text{with} \quad &w_1 a_{11} + w_2 a_{21} - w_3 = \mu_{c1}, \\ &w_1 a_{11} + w_2 a_{21} - w_3 = \mu_{c1}, \\ &w_1 a_{12} + w_2 a_{22} - w_4 = \mu_{c2}, \qquad 45 \\ &w_1 a_{12} + w_2 a_{22} - w_4 = \mu_{c2}, \\ &w_1, w_2, w_3, w_4 \geqslant 0. \end{aligned}$$

Eliminating the obvious redundancies this becomes

$$\begin{aligned} &\min w_1 \mu_{b_1} + w_2 \mu_{b_2}, \qquad 46 \\ \text{subject to} \quad &w_1 a_{11} + w_2 a_{21} - w_3 = \mu_{c_1}, \\ &w_1 a_{12} + w_2 a_{22} - w_4 = \mu_{c2}, \\ &w_1 w_2, w_3, w_4 \geqslant 0. \end{aligned}$$

But the dual to **46** is precisely the problem with which we began – viz., **41**.

We now specialize **41** to **40** and apply the rule **42** to achieve

$$\begin{aligned} & \max\ 12d_{11} + 10d_{12} + 6d_{21} + 5d_{22}, \\ \text{subject to}\quad & 36d_{11} + 30d_{12} + 24d_{21} + 20d_{22} + v_1 \leqslant 12, \\ & 60d_{11} + 50d_{12} + v_2 \leqslant 10, \\ & -12d_{11} - 10d_{12} + v_3 \leqslant 0, \\ & -12d_{21} - 10d_{22} + v_4 \leqslant 0, \\ & v_1, v_2, v_3, v_4 \geqslant 0. \end{aligned} \quad \mathbf{47}$$

A simplex calculation, as in Table I, produces an optimum at stage 2 with $d_{11}^* = \frac{1}{6}$, $d_{21}^* = \frac{1}{4}$ and all other $d_{ij}^* = 0$. Thus, applying **42** we have

$$\begin{aligned} x_1 &= \tfrac{1}{6} \times 12 = 2, \\ x_2 &= \tfrac{1}{4} \times 12 = 3. \end{aligned} \quad \mathbf{48}$$

An alternate optimum is shown at stage 3 with $d_{12}^* = \frac{1}{5}$, $d_{21}^* = \frac{1}{4}$ and all other $d_{ij}^* = 0$. Again applying **42** we have

$$\begin{aligned} x_1 &= \tfrac{1}{5} \times 10 = 2, \\ x_2 &= \tfrac{1}{4} \times 12 = 3. \end{aligned}$$

As is readily verified from **40** these x_j values are indeed optimal with an optimal return of $z^* = \$3{\cdot}50$. The programmed x_j^*'s obtained from **40** are in fact the same values as given for v_3^* and v_4^* in stages 2 and 3 of Table 1 and, furthermore, the $z_j^* - c_j$ values under V_1 and V_2 in these tableaus are the optimal solutions for the dual associated with **40**.[29]

We now extend the example of **40** by replacing our previous assumptions with the following ones

$$\sigma_{b1}^2 = 1, \qquad \sigma_{b2}^2 = 100, \qquad \operatorname{cov}(b_1, b_2) = 10, \quad \mathbf{49}$$

so that

$$\rho(b_1, b_2) = \frac{\operatorname{cov}(b_1, b_2)}{\sigma_{b1}\,\sigma_{b2}} = 1. \quad \mathbf{50}$$

In short, we assume that the variates b_1, b_2 are perfectly correlated with

29. See Charnes and Cooper (1961, ch. 1 ff.) for further discussion of this example as well as an explanation of the conventions used in Table 1.

$$\hat{b}_1 = b_1 - \mu_{b1} = b_1 - 12,$$
$$\hat{b}_2 = b_2 - \mu_{b2} = b_2 - 10,$$

and that $\hat{b}_1,\hat{b}_2$ are normally distributed about zero.

Table 1

		12	*10*	*6*	*5*				*Stage*
	P_0	P_{11}	P_{12}	P_{21}	P_{22}	V_1	V_2	V_3	V_4
V_1	12	36	30	24	20	1			
← V_2	10	60	50				1		
V_3	0	−12	−10					1	
V_4	0	↑		−12	−10				1
z_j-c_j	0	−12	−10	− 6	− 5				
← V_1	6		0	24	20	1	−36/60		
12 P_{11}	1/6	1	5/6				1/60		
V_3	2		0				12/60	1	
V_4	0			−12 ↑	−10				1
z_j-c_j	2		0	− 6	− 5		12/60		
6 P_{21}	1/4			1	5/6	1/24	− 1/40		
←12 P_{11}	1/6	1	5/6				1/60		
V_3	2						12/60	1	
V_4	3		↑			1/2	−12/40		1
z_j-c_j	3·5		0		0	1/4	1/20		
6 P_{21}	1/4			1	5/6	1/24	1/40		
10 P_{12}	1/5	6/5	1				1/50		
V_3	2						12/60	1	
V_4	3					1/2	−12/60		1
z_j-c_j	3·5	0			0	1/4	1/20		

These and the following data have been selected to give th

same result as before. But to make our illustration somewhat more interesting we may assume that our manufacturer is considering committing himself to supply his entire output to one customer. The latter is concerned only with knowing, with certainty, the proportions in which he will receive the two products and will accept any proportions offered, contracting to pay for them at the market price prevailing on the date of delivery. These prices are known to deviate normally about

$$\mu_{c_1} = \$1, \qquad \mu_{c_2} = \$\tfrac{1}{2}. \qquad \textbf{52}$$

With distributions such as normal densities one cannot require $x_1, x_2 \geqslant 0$ with certainty. We could relax these to probabilistic constraints for $a_i < 1$, but we shall instead merely waive the conditions for non-negativity entirely.

Suppose that our hypothetical manufacturer wishes to maximize his expected profits and subjects his constraints to $K_{\alpha i} = 2$, all i. We can, of course, utilize the previously developed constraints on the $\mu_i(D)$ since these are free of the $K_{\alpha i}$ risk components. To develop the remaining constraints we refer to the footnote given for **30** and write

$$\begin{aligned}\hat{\sigma}_1{}^2(D) &= E[a_1{}'\,D\hat{b} - \hat{b}_1]^2\\ &= E[\hat{b}_1 - (3d_{11} + 2d_{21})\hat{b}_1 - (3d_{12} + 2d_{22})\hat{b}_2]^2\\ &= \sigma_{b1}^2[(3d_{11} + 2d_{21}) - 1]^2 + \sigma_{b2}^2[3d_{12} + 2d_{22}]^2\\ &\quad + 2\rho(b_1, b_2)\sigma_{b1}\,\sigma_{b2}(3d_{12} + 2d_{22})[(3d_{11} + 2d_{21}) - 1].\end{aligned} \qquad \textbf{53}$$

Since, by **50**, we have $\rho = 1$ this becomes

$$\hat{\sigma}_1{}^2(D) = \{\sigma_{b1}[(3d_{11} + 2d_{21}) - 1] + \sigma_{b2}(3d_{12} + 2d_{22})\}^2. \qquad \textbf{54}$$

The corresponding constraint in **29** is

$$v_1{}^2 \geqslant K_{\alpha 1}^2\,\hat{\sigma}_1{}^2(D) = 4\hat{\sigma}_1{}^2(D). \qquad \textbf{55}$$

Thus, taking the positive square root, we have

$$v_1 \geqslant 2\{3d_{11} + 2d_{21} - 1 + 10(3d_{12} + 2d_{22})\}, \qquad \textbf{56}$$

or,

$$6d_{11} + 4d_{21} + 60d_{12} + 40d_{22} - v \leqslant 2. \qquad \textbf{57}$$

By an analogous development we also obtain

$$10d_{11} + 100d_{12} - v_2 \leqslant 20, \qquad \textbf{58}$$

where, of course, $v_1, v_2 \geqslant 0$ applies.

Collecting all details together and applying **29**:

$$\max 12d_{11} + 10d_{12} + 6d_{21} + 5d_{22}$$

$$\text{subject to} \quad \begin{aligned} 36d_{11} + 30d_{12} + 24d_{21} + 20d_{22} + v_1 &\leqslant 12, \\ 60d_{11} + 50d_{12} \qquad\qquad\qquad + v_2 &\leqslant 10, \\ 6d_{11} + 60d_{12} + 4d_{21} + 40d_{22} - v_1 &\leqslant 2, \\ 10d_{11} + 100d_{12} \qquad\qquad\qquad - v_2 &\leqslant 20, \\ v_1, v_2 &> 0. \end{aligned} \qquad \mathbf{59}$$

We have arranged the data of this example so that it gives the same results as before – e.g., $d_{11}^* = \frac{1}{6}$, $d_{12}^* = \frac{1}{4}$, as in **48**. The indicated assumptions also produce a dual for this case in the form

$$\min 12w_1 + 10w_2 + 2w_3 + 20w_4$$

$$\text{subject to} \quad \begin{aligned} 36w_1 + 60w_2 + 6w_3 + 10w_4 &= 12, \\ 30w_1 + 50w_2 + 60w_3 + 100w_4 &= 6, \\ 24w_1 \qquad + 4w_3 \qquad &= 6, \\ 20w_1 \qquad + 40w_3 \qquad &= 5, \\ w_1 \qquad - w_3 \qquad &\geqslant 0, \\ w_2 \qquad - w_4 &\geqslant 0. \end{aligned} \qquad \mathbf{60}$$

Observing the functional elements $\mu_{b_1} = 12$, $\mu_{b_2} = 10$, and $K_{a_1}\sigma_{b_1} = 2$, $K_{a_2}\sigma_{b_2} = 20$ and then observing the last pair of constraints – which arise from v_1, v_2 pairs in the direct problem – we can see that values for the dual variables associated with the risk terms can never exceed those associated with improvements in the means μ_{b1} and μ_{b2}.

References

BEN-ISRAEL, A. (1962), 'On some problems of mathematical programming', Ph.D. Thesis in Engineering Science, Northwestern University, Illinois.

BEN-ISRAEL, A. and CHARNES, A. (1961), 'Constant level inventory policies and chance-constrained programming models', Northwestern University, Illinois.

CHARNES, A., and COOPER, W. W. (1958), 'The theory of search: optimum distribution of search effort', *Management Science*, vol. 5, no. 1.

CHARNES, A., and COOPER, W. W. (1959), 'Chance-constrained programming', *Management Science*, vol. 6, no. 1.

CHARNES, A., and COOPER, W. W. (1961), *Management Models and Industrial Applications of Linear Programming*, Wiley.

CHARNES, A., and COOPER, W. W. (1962a), 'Normal deviates and chance constraints', *J. Am. Stat. Assoc.*, vol. 57, pp. 134–48.

CHARNES, A., and COOPER, W. W. (1962b), 'Programming with fractional functionals, I: linear fractional programming', *Naval Res. Logistics Q.*, vol. 9, pp. 181-6.

CHARNES, A., and COOPER, W. W. (1962c), 'Systems evaluations and repricing theorems', *Management Science*, vol. 9, no. 1.

CHARNES, A., COOPER, W. W., and KORTANEK, K. (1962), 'Duality, Haar programs and finite sequence spaces', *Proc. Nat. Acad. Sci.*, vol. 48, pp. 783–6.

CHARNES, A., COOPER, W. W., and KORTANEK, K. (1963) 'Duality in semi-infinite programs and some works of Haar and Caratheodory', *Management Science*, vol. 9, pp, 209-28.

CHARNES, A., COOPER, W. W., and MILLER, M. H. (1959), 'Application of linear programming to financial budgeting and the costing of funds', *J. Business Univ. of Chicago*, vol. 32, pp. 20–46.

CHARNES, A., COOPER, W. W., and SYMONDS, G. H. (1958), 'Cost horizons and certainty equivalents: an approach to stochastic programming of heating oil production', *Management Science*, vol. 4, no. 3.

CHARNES, A., COOPER, W. W., and THOMPSON, G. L., (1965), 'Constrained generalized medians and linear programming under uncertainty', *Management Science*, vol. 12, pp. 83-112.

COOPER, W. W., and SAVVAS, J. D. (1961), 'Motivational cost and transients in budgeting the behavior of cost and aspirations', ONR Memo 83, Carnegie Institute of Technology.

DANTZIG, G. B. (1955), 'Linear programming under uncertainty', *Management Science*, vol. 1, pp. 197–206.

DANTZIG, G. B. (1956), 'Recent advances in linear programming', *Management Science*, vol. 2, pp. 131–44.

HOLT, C. C., MODIGLIANI, F., MUTH, J., and SIMON, H. A. (1960), *Planning Production Inventory and Work Force*, Prentice-Hall.

MANDANSKY, A. (1960), 'Inequalities for stochastic linear programming problems', *Management Science*, vol. 6, pp. 197–204.

MARCH, J. G., and SIMON, H. A. (1958), *Organizations*, Wiley.

MARKOWITZ, H. M. (1959), *Portfolio Selections: Efficient Diversification of Investment*, Wiley.

NASLUND, B., and WHINSTON, A. W. (1962), 'A model of multi-period investment under uncertainty', *Management Science*, vol. 8, pp. 184–200.

SIMON, H. A. (1957), *Models of Man*, Wiley.

SIMON, H. A. (1959), 'Theories of decision-making in economics and behavioral science', *Am. Econ. Rev.*, vol. 49, no. 3.

STEDRY, A. (1960), *Budget Control and Cost Behavior*, Prentice-Hall.

STEDRY, A., and CHARNES, A. (1962), 'Exploratory models in the theory of budgetary control', ONR Research Memo 43, Mimeo, for project *Temporal Planning and Management Decisions under Risk and*

Uncertainty, Northwestern University, Ill.

SYMONDS, G. (1962), 'Stochastic scheduling by the horizon method', *Management Science*, vol. 8, pp. 138–67.

THEIL, H. (1961), 'Some reflections on static programming under uncertainty', *Weltwirtschaftliches Archiv*, vol. 87, and reprinted as Publication no. 5 of The International Center for Management Science, The Netherlands School of Economics.

TINTNER, G. (1955), 'Stochastic linear programming with applications to agricultural economics', in H. A. Antosiewicz (ed.) *Second Symposium on Linear Programming*, National Bureau of Standards, Washington.

TINTNER, G. (1960), 'A note on stochastic linear programming', *Econometrica*, vol. 28, pp. 490–95.

VAJDA, S. (1958), 'Inequalities in stochastic linear programming', *Bull. Int. Stat. Inst.*, vol. 36, pp. 357–63.

VAN DE PANNE, C., and POPP, W. (1963), 'Minimum-cost cattle feed under probabilistic protein constraints', *Management Science*, vol. 9, pp. 405-30

WEINGARTNER, H. M. (1961), 'Mathematical programming and the analysis of capital budgeting', Ph.D. Thesis, Graduate School of Industrial Administration, Carnegie Institute of Technology.

15 S. E. Elmaghraby

An Approach to Linear Programming under Uncertainty

S. E. Elmaghraby, 'An approach to linear programming under uncertainty', *Operations Research*, vol. 7, 1959, pp. 208–16.

A characteristic of linear-programming problems is the deterministic nature of the parameters involved. For example, the allocation of capital resources is always to a fixed and known demand, the vitamin content of various diets (in diet problems) is known deterministically with no possible variation, etc. Clearly, such an ideal state of knowledge is scarcely, if ever, realized. In many situations, production is to a *varying* demand, while almost all of the parameters (representing cost, yield, effectiveness, etc.) exhibit stochastic variation.

This variability (as measured by the variance of the chance variable, say) may be insignificant. In such a case, the variability may be ignored, and the results obtained by linear-programming techniques (which implicitly assume the constant character of the parameters) are satisfactory for all practical purposes. However, there are many instances of programming problems in which the variability in some (or all) of the parameters is considerable and cannot be ignored. Clearly, the results obtained by assuming that the parameters are constant (at some average value) must be qualified and, if unreservedly adopted, may prove to be a source of serious errors and subsequent economic losses.

In the presence of stochastic variation, the current practice, as was pointed out by Dantzig (1956), is to ignore the random variation and provide plenty of 'fat' in the deterministic version of the problem, in the hope that the uncertainty of the future will be completely absorbed by such 'play', and the program will be executed without failure. As long as capabilities are well above requirements, or if the demands could be shifted in time, this approach may be adequate. However, where the capabilities are limited, the above procedure can lead to actions far from the optimal.

The problem of programming under uncertainty has been shied away from, mostly because of the fact that the formulation of the problem usually leads to objective functions of the nonlinear type. *Prima facie*, this eliminates the possibility of utilizing the powerful techniques of linear programming. However, closer study of certain problems showed that by utilizing simple tricks, linear-programming techniques can be used to obtain (or at least approximate) the required solution. An example is Charnes' and Lemke's (1954) treatment of programming problems with nonlinear separable convex objective functions,[1] which shows that an approximation of the solution can be carried out to any desired degree of accuracy.

In the following we present a solution to a programming problem under uncertainty when the distribution function of demand is *discrete* and the cost structure assumes a certain type. The resulting objective function is convex, but not necessarily separable. Still, with a rather simple change in variables, the solution of the stochastic problem is shown to be equivalent to the solution of a linear-programming problem. We note that no approximation is involved in this procedure.

The statement of the problem as well as the numerical example presented in the text yield a separable convex objective function. However, the Appendix contains the formulation of a problem that yields a convex (but not separable) objective function. The theory presented below applies to both types of objective functions.

Assumptions

For clarity of exposition, we shall formulate the programming problem under stochastic demand in the production and inventory language. We shall assume that:

1. The demand for the product(s) is the only chance variable, and all other factors and parameters are known deterministically.

1. A separable convex function of the variables X ,..., X_n is a function that, with suitable linear transformation of variables can be exhibited in the form: $f(X_i,\ldots,X_n) = \Sigma f_j(X_j), j = 1,\ldots, n$, where $f_j(X_j)$ is convex in X_j.

2. The costs of production are composed of: (a) costs of providing the elements of production and maintaining them in working conditions, (b) costs of storage of unsold products, and (c) costs of denying sales to a customer because of depletion of stocks (this latter cost is usually an opportunity cost).

3. The demands for the different products are independent.

4. The demand distribution function is stationary in time.

Notation

$p(d_{jh})$ = the probability frequency function (p.f.f.) of the demand d_{jh}, for the jth product, $j = 1, \ldots, n$, $h = 1, \ldots, s_j$; clearly

$$d_{jh} \geqslant 0,\ p(d_{jh}) \geqslant 0,\ \sum_{h=1}^{h=s_j} p(d_{jh}) = 1.$$

c_j = the cost per unit of time of 'manufacture' of a unit of product j.

C_{0j} = the cost of storage per unit time of a unit of item j.

C_{uj} = the cost of shortage per unit time of a unit of item j.

a_{ij} = constants representing the coefficients of the restrictions imposed on the production system $i = 1, \ldots, m$; $j = 1, \ldots, n$; these constants involve availability as well as technological restrictions.

b_i = the limits of the available capabilities, $i = 1, \ldots, m$.

X_j = the (unknown) amount produced of item j per unit time.

Statement of Problem

The problem of programming under stochastic demand can be formulated as follows: Minimize (or maximize)

$$C(X_j) \qquad (j = 1, \ldots, n) \qquad 1$$

subject to the linear restrictions

$$\sum_j a_{ij} X_j = b_i, \qquad (i = 1, \ldots, m) \qquad 2$$

where C is a convex (separable or nonseparable) function in the X's. Under assumption 2, and assuming that we wish to minimize the total expected cost, it is easy to see that the objective would be to minimize

$$C(X_j) = \sum_{j=1}^{j=n} c_j X_j + \sum_j C_{0j} \sum_{d_{jh}=0}^{X_j} (X_j - d_{jh})\, p(d_{jh}) +$$

$$+ \sum_j C_{uj} \sum_{d_{jh}=X_j}^{\infty} + (d_{jh} - X_j)\, p(d_{jh}), \qquad 3$$

subject to restrictions **2**.

Representation of the Stochastic Programming Problem as a Linear-Programming Problem

Let the points of definition of the p.f.f. of demand for the jth product be $d_{j1}, d_{j2}, \ldots, d_{js_j}$, as in Figure 1. We have

$$p(d_{jh}) \geqslant 0, \qquad \sum_{h=1}^{h=sj} p(d_{jh}) = 1, \qquad d_{js_j} < \infty .$$

Since X_j is the total amount produced of item j, $0 \leqslant X_j < \infty$; then X_j must satisfy one of the following inequalities:

$$0 \leqslant X_j < d_{j1},\ d_{j1} \leqslant X_j < d_{j2}, \ldots,\ d_j(s_j - 1) \leqslant X_j < d_{js_j}, \qquad 4$$
$$d_{js_j} \leqslant X_j < \infty .$$

Let X_{jh} denote the fact that X_j satisfies the hth inequality for product j, where $h = 1, \ldots, s_j + 1$. Thus X_{jh} denotes that

$$d_{j(h-1)} \leqslant X_j < d_{jh}. \qquad 5$$

The partial difference of the objective function **3** with respect to X_{jh} (noting relation **5**) is given by

$$D_{jh} = \frac{\Delta C}{\Delta X_{jh}} = c_j + C_{0j} \sum_{d_{jh}=0}^{X_j} p(d_{jh}) - C_{uj} \sum_{X_i}^{\infty} + p(d_{jh}), \qquad 6[2]$$

which is *constant* for any X_{jh}, and is *increasing* in h. Let

$$X_j = \sum_k X_{jk}, \qquad (k = 1, \ldots, s_j + 1) \qquad 7$$

i.e. we shall consider the total production of the jth product as the sum of smaller amounts $\{X_{jk}\}$.[3] Clearly $C(X_j)$ can be

2. The notation X_j^+ means: 'just greater than'.
3. The notation $\{\ \}$ means: the 'set of'.

written in terms of $\{X_{jk}\}$ by substituting **7** in **1**. It is easy to show that the function (X_j) is convex in the region of definition of $\{X_j\}$ (Fenchel, 1953).

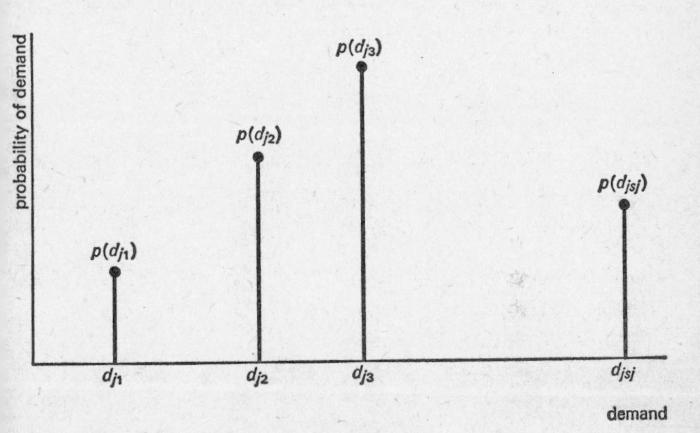

Figure 1

Next, consider the (triangular) array of restrictions

$$\sum_{k=1}^{k=h} X_{jk} \leqslant d_{jh}, \qquad (h = 1, \ldots, s_j + 1) \qquad \mathbf{8}$$

where $X_{jk} \geqslant 0$. We have the following

Theorem: If $\{X^0_{jk}\}$ is the solution of the linear program defined as follows:

maximize

$$F(X_{jk}) = \sum_j \sum_k (-D_{jk} X_{jk}),$$

where $D_{jk} = D_{jh}$ for subscript $k = h$, subject to the linear restrictions **2** *with $\sum_k X_{jk}$ substituted for X_j $(j = 1, \ldots, n; k = 1, \ldots, s_j + 1)$ and the restrictions* **8**, *then*

$$\{X^0_j\} = \sum_{k=1}^{sj+1} X^0_{jk}$$

is the solution to the original programming problem.

Proof. The cost function $C(X_j)$ is piecewise *linear* in $\{X_j\}$, since its partial difference, D_{jh} (and consequently D_{jk}), is constant for the interval of definition of X_{jh} (*see* **6** above). Also, by the convexity of $C(X_j)$ in $\{X_j\}$, the partial differences are nondecreasing (in fact, strictly increasing). Therefore, $D_{jk} = D_{jh} \leqslant D_{j,h+1} = D_{j,\,k+1}$ for $k = h$; $D_{jk} = D_{jh} =$ a constant for $d_{j,h-1} \leqslant X_{jh} < d_{jh}$. Suppose that the allocation $\{X_j'\}$ yields exactly $X_1' = d_{1h_1}$, $X_2' = d_{2h_2}$, ..., $X_j' = d_{jh_j}$, ..., $X_n' = d_{nhn}$, then for *some* increase in the $\{X_j'\}$'s, the amounts produced will lie in the region of definition of the respective $\{X_{jhj+1}\}$.

By the fact that the cost function **3** is linear in $\{X_j\}$ within that interval, the question of which X_j to increase and which to maintain constant at its value X_j' is answered by *the linear program*: Minimize $\sum_j D_j X_j$ plus a constant depending on

$(d_{1h_1}, d_{2h_2}, \ldots, d_{nh_n})$, subject to $\sum_j a_{ij} X_j = b_i - \sum_j a_{ij} X_j'$.

Clearly, this linear programme is applicable only within the region of definition of $\{X'_{jhj+1}\}$ (which is the region in which D_j is constant and equal to D_{jhj+1}), i.e., applicable in the region where $d_{jh_j} \leqslant X_j \leqslant d_{jh_j+1}$ for $j = 1, \ldots, n$; $h = 1, \ldots, s_j$. When the boundary of the region is reached *in any j*, the partial differences change values, and the process of obtaining the next increase in the new allocation is started in the same fashion as above, but with the new $d_{j,h+1}$'s and the new D_j's.

It is, therefore, clear that the solution of the *original* problem is obtained through a *sequence* of linear programs, each affecting the X_j's within the interval of definition of the X_{jh}'s.

A study of the (triangular) restrictions **8** reveals that the notation X_{jk} stands for the increment of X_j within the interval of definition of X_{jh}. In addition, since D_{jk} (which is the *same* as D_{jh} for $d_{j,h-1} \leqslant X_j \leqslant d_{jh}$) has a nondecreasing character, $X_{j,k+1}$ will never be > 0 before X_{jk} has been fully utilized. The reason for this is as follows: in every linear restriction in which $X_{j,\,k+1}$ appears, X_{jk} also appears *with the same coefficient*. Hence, the quantity z_j (in conventional simplex notation – see Charnes, 1953) is the same for both X_{jk} and $X_{j,k+1}$. The partial difference D_{jk} represents the cost coefficient c_j in conventional simplex notation. Since $D_{jk} \leqslant D_{j,\,k+1}$, it follows that $z_j -$

$D_{jk} \geqslant z_j - D_{j,\,k+1}$. Remembering that the allowable increase in X_{jk} is only $d_{j,\,h+1} - d_{jh}$ while the allowable increase in $X_{j,\,k+1}$ is $d_{j,\,h+2} - d_{jh}$, and since $d_{j,\,h+2} - d_{jh} > d_{j,\,h+1} - d_{jh}$, X_{jk} will be chosen as a basis vector *before* $X_{j,\,k+1}$, and the theorem follows. The *value* of the objective function **3** at the optimal allocation is to be evaluated directly.

A Numerical Example

The following example is presented in order to illustrate the procedure outlined above. The example is hypothetical and is deliberately simplified in order not to obscure the few basic notions of the procedure.

A machine shop has one drill press and five milling machines, which are to be used to produce two parts, 1 and 2. The relevant data on production rates, costs, and selling prices are given in the table below:

Part	*Production time (min/piece)*		*Selling price ($/piece)*	C_{oj} *($/piece)*	C_{uj} *($/piece)*
	Drill	*Mill*			
1	3	20	1·35	0·25	1·40
2	5	15	1·50	0·18	1·55
Operating costs ($/hr)	1·80	2·40			

It is desired to maintain a balanced loading on all machines so that no machine runs more than sixty minutes per day longer than any other. The maximum working hours per day are eight. The demand per period for the two products is as follows:

Part	*Demand* d_{jh}	*Probability* $p(d_{jh})$
1	50	0·20
	60	0·40
	70	0·40
2	20	0·10
	25	0·35
	30	0·35
	35	0·20

Required: the amount to be produced of each item to maximize *expected* profit.
Let

X_1 = number of pieces produced of part 1.
X_2 = number of pieces produced of part 2.

Time Restrictions

Drill: $3\ X_1 + 5\ X_2 \leqslant 1 \times 8 \times 60 = 480,$ **9a**

Mill: $20\ X_1 + 15\ X_2 \leqslant 5 \times 8 \times 60,$

$4\ X_1 + 3\ X_2 \leqslant 480.$ **9b**

Load-Balance Restrictions

$$|(3\ X_1 + 5\ X_2) - \tfrac{1}{5}(20\ X_1 + 15\ X_2)| \leqslant 60,$$

i.e.,

$X_1 - 2\ X_2 \leqslant 60,$ **10a**

$2\ X_2 - X_1 \leqslant 60.$ **10b**

Criterion Function

The expected profits in dollars per period are given by

$$P = 1{\cdot}35[\sum_{d_{1h}=0}^{X1} d_{1h}\ p(d_{1h}) + X_1 \sum_{d_{1h}=X_1}^{\infty} + p(d_{1h})] + 1{\cdot}50[\sum_{d_{2h}=0}^{X2} d_{2h}$$

$$p(d_{2h}) + X_2 \sum_{d_{2h}=X_2}^{\infty} + p(d_{2h})] - \tfrac{1}{60}(3\ X_1 + 5\ X_2) \times 1{\cdot}80 -$$

$$- \tfrac{1}{60}(20\ X_1 + 15\ X_2) \times 2{\cdot}40$$

$$- 0{\cdot}25 \sum_{d_{1h}=0}^{X1} (X_1 - d_{1h})\ p(d_{1h}) - 0{\cdot}18 \sum_{d_{2h}=0}^{X2} (X_2 - d_{2h})\ p(d_{2h})$$

$$- 1{\cdot}40 \sum_{d_{1h}=X_1}^{\infty} + (d_{1h} - X_1)\ p(d_{1h}) - 1{\cdot}55 \sum_{d_{2h}=X_2}^{\infty} + (d_{2h} - X_2)\ p(d_{2h})$$

$$= - 0{\cdot}89\ X_1 - 0{\cdot}75\ X_2 + 1{\cdot}35\ [\sum_{0}^{X_1} d_{1h}\ p(d_{1h}) + X_1 \sum_{X_1}^{\infty} + p(d_{1h})]$$

$$+1{\cdot}50\,[\sum_{0}^{X_2} d_{2h}\,p(d_{2h}) + X_2 \sum_{X_2}^{\infty} +\, p(d_{2h})] -$$

$$-\,0{\cdot}25 \sum_{0}^{X_1} (X_1 - d_{1h})\,p(d_{1h})$$

$$-\,0{\cdot}18 \sum_{0}^{X_2} (X_2 - d_{2h})\,p(d_{2h}) - 1{\cdot}40 \sum_{X_1}^{\infty} + (d_{1h} - X_1)\,p(d_{1h})$$

$$-\,1{\cdot}55 \sum_{X_2}^{\infty} + (d_{2h} - X_2)\,p(d_{2h},)$$

which is the function to be maximized.

To transform the problem into a linear program, we evaluate the partial difference of the criterion function (which is defined for all X_j's except the points of definition of the p.f.f.). It is well to note that the partial differences are identical with the marginal profit for any X_1 and X_2 (other than the points d_{jh}). Thus,

$$\Delta P/\Delta X_1 = 2{\cdot}75 \sum_{X1}^{\infty} + p(d_{1h}) - 0{\cdot}25 \sum_{0}^{X_1} p(d_{1h}) - 0{\cdot}89,$$

$$\text{and } \Delta P/\Delta X_2 = 2{\cdot}45 \sum_{X_2}^{\infty} + p(d_{2h}) - 0{\cdot}18 \sum_{0}^{X_2} p(d_{2h}) - 0{\cdot}75.$$

It is evident that for any given X_1 and X_2 (other than the points $d_{jh}, j = 1, 2; h = 1, \ldots, s_j; s_1 = 3$ for $j = 1$, and $s_2 = 4$ for $j = 2$) the marginal profits are constant, and their values change only *at* the points d_{jh}. The following schedule can then be calculated.

Schedule of Marginal Revenues, $\{D_{jh}\}$

Range of X_1	$\Delta P/\Delta X_1$	*Range of* X_2	$\Delta P/\Delta X_2$
$0 \leqslant X_1 < 50$	1·86	$0 \leqslant X_2 < 20$	1·70
$50 \leqslant X_1 < 60$	1·26	$20 \leqslant X_2 < 25$	1·437
$60 \leqslant X_1 < 70$	0·06	$25 \leqslant X_2 < 30$	0·5165
		$30 \leqslant X_2 < 35$	−0·404

Define the variables X_{11}, X_{12}, X_{13}; and X_{21}, X_{22}, X_{23} X_{24} to satisfy the restrictions

$$\sum_{k_1=1}^{3} X_{1k_1} = X_1, \qquad \textbf{12a}$$

$$\sum_{k1=1}^{h1} X_{1k1} = \leqslant d_{1h1}, \qquad (h_1 = 1, 2, 3) \quad \textbf{12b}$$

$$\sum_{k2=1}^{4} X_{2k} = X_2, \qquad \textbf{13a}$$

$$\sum_{k_2=1}^{h2} X_{2k_2} \leqslant d_{2h_2}. \qquad (h_2 = 1, 2, 3, 4) \quad \textbf{13b}$$

The original (stochastic) problem is now equivalent to the linear programming problem defined as follows: *Minimize*

$$G(X_{jk}) = -0{\cdot}86\ X_{11} - 1{\cdot}26\ X_{12} - 0{\cdot}06\ X_{13} - 1{\cdot}70\ X_{21} - 1{\cdot}437\ X_{22} - 0{\cdot}5165\ X_{23} + 0{\cdot}404\ X_{24}, \qquad \textbf{14}$$

$$3 \sum_{k1=1}^{3} X_{1k_1} + 5 \sum_{k2=1}^{4} X_{2k_2} \leqslant 480, \qquad \textbf{9a}'$$

$$4 \sum_{k_1=1}^{3} X_{1m} + 3 \sum_{k_2=1}^{4} X_{2k_2} \leqslant 480, \qquad \textbf{9b}'$$

$$\sum_{k_1=1}^{3} X_{1k_1} - 2 \sum_{k_2=1}^{4} X_{2k_2} \leqslant 60, \qquad \textbf{10a}'$$

$$2 \sum_{k_2=1}^{4} X_{2k_2} - \sum_{k_1=1}^{3} X_{1k_1} \leqslant 60, \qquad \textbf{10b}'$$

$$\sum_{k_1=1}^{h_1} X_{1k_1} \leqslant d_{1h_1}, \qquad (h_1 = 1, 2, 3) \quad \textbf{12b}'$$

$$\sum_{k_2=1}^{h_2} X_{2k_2} \leqslant d_{2h_2}. \qquad (h_2 = 1, 2, 3, 4) \quad \textbf{13b}'$$

Thus, while the original (stochastic) problem contained four equations in six unknowns (two basic and four slack), the new formulation (as a linear program) contains eleven equations (the original four plus $s_1 + s_2 = 7$) in 18 unknowns (seven basic, eleven slack).

The total expected profit from any allocation (characterized by a pair of numbers X_1, X_2) is easily calculated from the criterion function **11**. Since the linear-programming iterations usually start at the values $X_{1k_1} = 0$, $X_{2k_2} = 0$, for all k_1 and k_2 this will imply (by **12a** and **13a**) that $X_1 = 0$ and $X_2 = 0$. Let the solution of the linear-programming problem be the vectors $\{X^0_{1k_1}\}$ and $\{X^0_{2k_2}\}$ with a corresponding value function $G(\{X^0_{1k_1}\}, \{X^0_{2k_2}\})$. Let (X^0_1, X^0_2) be the production schedule referred back to the original problem by **12a** and **13a**. The expected profit from such a production schedule then can be evaluated directly by inserting the values (X^0_1, X^0_2) in **11**.

The solution of the linear-programming problem follows conventional methods and shall not be pursued here any further. We pause only to state that the solution of the above numerical example yields $X^0_1 = 70$ units, and $X^0_2 = 30$ units, at a total expected profit of $35.73. If no stochastic variation were involved and it were required to maximize profit under the assumption that the demand for both products was unlimited, the optimal schedule of production would have been $X_1 = 87$ units and $X_2 = 44$ units and the total profit = $149.82.

Remarks

It is noted that the above procedure suffers from two limitations: (a) It is applicable only to discrete distribution functions. The reason for such limitation is that the demand function should have a *finite* number of points of definition, $\{d_{jh}\}$, in order that upper bounds of the form of **12b** and **13b** can be applied. A continuous demand distribution function will have to be approximated by a discontinuous one and, since the size of the problem (which is measured by the number of equations present) is directly proportional to the number of points $\{d_{jh}\}$, it is easy to see that too fine an approximation may yield a problem too large even for the latest electronic computer. (b) The criterion function possesses a *constant* marginal value, except at the points $\{d_{jh}\}$. It is this limitation that curbs the power of the technique and may limit it to a relatively small class of programming problems.

However, it is important to note the fact that the increase in the number of restrictions is by only $\sum_j s_j$, which is a relatively small increase compared to the ease of solution achieved through the reduction of the stochastic problem to a linear-programming one.

Appendix

In the following we present a problem of stochastic programming that leads to convex but not separable objective functions. Let

X_{ij} = the amount of resource i devoted to product j in one period, $i = 1, \ldots, mj = 1, \ldots, n$
r_{ij} = the productivity (in units of product per unit time) of resource i when devoted to product j
c_{ij} = the cost of manufacture of a unit of product j
S_j = total amount produced (per unit time) of product j

and $d_{jh}, p(d_{jh}), a_{ij}, b_i, C_{0j}, C_{uj}$ be defined as before. Clearly,

$$S_j = \sum_i r_{ij} X_{ij}.$$

The problem of programming under stochastic demand is: *Minimize*

$$C(X_{ij}) = \sum_i \sum_j c_{ij} r_{ij} X_{ij} + \sum_j C_{0j} \sum_{d_{jh}=0}^{S} (S_j - d_{jh})\, p(d_{jh})$$

$$+ \sum_j C_{uj} \sum_{d_{jh}=S_j}^{\infty} + (d_{jh} - S_j)\, p(d_jh)$$

$$= \sum_j [\sum_i c_{ij} r_{ij} X_{ij} + C_{0j} \sum_{d_{jh}=0}^{Sj} (S_j - d_{jh})\, p(d_{jh})$$

$$+ C_{uj} \sum_{d_{jh}=S_j}^{\infty} + (d_{jh} - S_j)\, p(d_{jh})],$$

subject to the linear restrictions $\sum_j a_{ij} X_{ij} \leqslant b_i$.

It can be shown that the quantity inside the square brackets in the objective function is convex, though not separable.

References

CHARNES, A., COOPER, W. W., and HENDERSON, A. (1953), *Introduction to Linear Programming*, Wiley.

CHARNES, A., and LEMKE, C. E. (1954), 'Minimization of non-linear separable convex functional', *Naval Res. Logistics Q.*, vol. 1, pp. 301–12.

DANTZIG, G. B. (1956), 'Recent advances in linear programming', *Management Science*, vol. 2, pp. 131–44.

FENCHEL, W. (1953), *Convex Cones, Sets, and Functions*, Princeton University Press, p. 87.

16 K. B. Haley and A. J. Smith

Transportation Problems with Additional Restrictions

K. B. Haley and A. J. Smith, 'Transportation problems with additional restrictions', *Applied Statistics*, no. 2, 1966, pp. 116–23.

Introduction

Problems which are formulated as transportation problems are often stated as: 'A concern which produces a product at a number of sources (factories) wishes to supply known amounts of the product to a number of destinations (customers) at minimum cost.' In these problems it is assumed that the product is identical no matter which source produces it and that the customers have no preference relating to the source of supply. However, in many practical instances, the product does vary in some characteristic according to its source. The final product mixture, received by the destinations, may then be required to meet known specifications.

In this paper a general problem of this type is described and a numerical illustration referring to the transport and processing of iron ore is considered in detail. This illustration is known to be artificial and it is assumed that the crude ore contains different amounts of an impurity, phosphorus, according to its source. The actual time to process the ore depends on both its source and destination. The problem is formulated as a transportation problem with two sets of additional restrictions imposed upon it.

A similar problem regarding the movement of coal is described by Williams and Haley (1959). Their general problem is a transportation problem with three sets of additional restrictions.

General Problem

A general form of this problem is to allocate a product which is available at N sources ($i = 1 \ldots N$) in quantities a_i to M destinations ($j = 1 \ldots M$) which require b_j. One unit of the product contains p_{ijk} units of P impurities ($k = 1 \ldots P$) when it is sent from i to j. Customer j cannot receive more than d_{jk} units of impurity k. The cost of transporting one unit of the product from i to j is c_{ij} and it is required to satisfy the demand at minimum cost.

In the mathematical formulation let x_{ij} units be the amount sent from i to j then the problem becomes:

Minimize $$\sum_{i=1}^{N} \sum_{j=1}^{M} x_{ij}\, c_{ij} = C \qquad 1$$

subject to $$\sum_{i=1}^{N} x_{ij} = b_j \quad (j = 1 \ldots M), \qquad 2$$

$$\sum_{j=1}^{M} x_{ij} = a_i \quad (i = 1 \ldots N), \qquad 3$$

$$\sum_{i=1}^{N} p_{ijk}\, x_{ij} \leqslant d_{jk} \quad (j = 1 \ldots M;\ k = 1 \ldots P), \qquad 4$$

$$x_{ij} \geqslant 0, \qquad 5$$

$$\sum_{i=1}^{N} a_i = \sum_{j=1}^{M} b_j.$$

The restrictions **2**, **3** and **5** define an ordinary transport problem and there are P sets of additional restrictions **4**.

Method of Solution

This problem could be solved in a great variety of ways. We recommend the use of the (primal) simplex method, in the inverse matrix form. In fact the basis has a specially simple structure, which is used to calculate the relevant elements of its inverse from first principles, rather than updating them in the usual way.

The method of solution of this problem is an adaptation of the transportation technique analogous to the Routing Aircraft Problem (see Vajda, 1962). The structure of the equations is used to define a basic feasible solution and fictitious costs. A method is given to alter one basic feasible solution and produce another by the substitution of one non-basic variable for a basic variable. The necessary theory appears in the next sections and the method is applied to a small numerical example. A more detailed version of the proofs (relating to a different type of problem) is described by one of the authors elsewhere (see Haley. 1963).

Following the usual method of solution the first stage is to introduce slack variables x_{N+k-j} into the restrictions **4** to make them into equations, viz.

$$\sum_{i=1}^{N} p_{ijk}\, x_{ij} + x_{N+k-j} = d_{jk} \quad (j = 1 \ldots M;\ k = 1 \ldots P), \qquad \mathbf{6}$$

$$x_{N+k-j} \geqslant 0. \qquad \mathbf{7}$$

There are a total of $MN + MP$ variables including slacks and $MP + M + N$ equations. Because of the conditions imposed on the a_i and b_i one of the equations **2** and **3** is dependent and so a basic feasible solution contains $MP + M + N - 1$ basic variables.

Fictitious costs (which are dual variables) r_j, s_i, t_{jk} are defined so that

$$r_j + s_i + \sum_{k=1}^{P} t_{jk}\, p_{ijk} = c_{ij} \qquad \mathbf{8}$$

(for those i, j for which x_{ij} is in basis)
and $$t_{jk} = 0$$
(for those k, j for which x_{N+k-j} is in basis).
In an optimum solution it is shown that

$$r_j + s_i + \sum_{k=1}^{P} t_{jk}\, p_{ijk} \leqslant c_{ij}, \qquad \mathbf{10}$$

$$t_{jk} \leqslant 0. \qquad \mathbf{11}$$

An iterative procedure is carried out to obtain the solution, i.e. (i) A basic feasible solution is found and tested for optimality, (ii) Any improvement is made and a new basic feasible solution is found, (iii) The new solution is tested.

Fictitious Costs

The dual of the problem defined by **1, 2, 3, 5, 6, 7** is taken to give:

Maximize $$B = \sum_{j=1}^{M} r_j b_j + \sum_{i=1}^{N} s_i a_i + \sum_{j=1}^{M}\sum_{k=1}^{P} t_{jk} d_{jk} \qquad \textbf{12}$$

subject to $$r_j + s_i + \sum_{k=1}^{P} p_{ijk} t_{jk} \leqslant c_{ij}, \qquad \textbf{13}$$

$$t_{jk} \leqslant 0. \qquad \textbf{14}$$

In an optimum solution the values of the objective functions of the primal and dual are equal (Gass, 1958). To find a condition for optimality these objective functions C and B are equated giving

$$\sum_{i=1}^{N}\sum_{j=1}^{M} x_{ij} c_{ij} - \sum_{j=1}^{M} r_j b_j - \sum_{i=1}^{N} s_i a_i - \sum_{k=1}^{P}\sum_{j=1}^{M} t_{jk} d_{jk} = 0. \qquad \textbf{15}$$

Using equations **2, 3** and **6** equation **15** becomes:

$$\sum_{i=1}^{N}\sum_{j=1}^{M} x_{ij} c_{ij} - \sum_{j=1}^{M} r_j \sum_{i=1}^{N} x_{ij} - \sum_{i=1}^{N} s_i \sum_{j=1}^{M} x_{ij}$$

$$- \sum_{j=1}^{M}\sum_{k=1}^{P} t_{jk}\left(\sum_{i=1}^{N} p_{ijk} x_{ij} + x_{N+k-j} \right) = 0, \qquad \textbf{16}$$

$$\sum_{i=1}^{N}\sum_{j=1}^{M} \left(c_{ij} - r_j - s_i - \sum_{k=1}^{P} t_{jk} p_{ijk} \right) x_{ij} + \sum_{j=1}^{M}\sum_{k=1}^{P} (-t_{jk}) x_{N+k-j} = 0 \qquad \textbf{17}$$

From the conditions **5, 7, 12, 13** each term in equation **17** is greater than or equal to zero and hence the only way in which equation **17** can be satisfied is that each of the product terms are zero. That is, the conditions for optimality are:

$$c_{ij} = r_j + s_i + \sum_{k=1}^{F} t_{jk} p_{ijk} \qquad \textbf{18}$$

(for those i, j for which x_{ij} is in the basis)
and $$t_{jk} = 0 \qquad \textbf{19}$$
(for those j, k for which x_{N+k-j} is in basis).

Altering a Basic Feasible Solution

If a basic feasible solution is to be altered by the introduction of a non-basic variable and the removal of a basic one then alterations can only be made to basic variables. It is assumed, without loss of generality, that x_{11} is non-basic and that an unknown quantity θ is to be added to this variable. Let the alterations in the basic variables x_{rs}, x_{N+ts} be given by the addition of θn_{rs}, θn_{N+ts} to them. Then if the new solution satisfies the original constraints the n's must satisfy

$$\sum_{r=1}^{N} n_{rs} = 0 \quad (s = 1 \ldots M), \qquad \textbf{20}$$

$$\sum_{s=1}^{M} n_{rs} = 0 \quad (r = 1 \ldots N), \qquad \textbf{21}$$

$$\sum_{r=1}^{P} p_{rst} n_{rs} + n_{N+ts} = 0 \quad (s = 1 \ldots M; t = 1 \ldots P), \qquad \textbf{22}$$

where $n_{11} = 1$; $n_{rs} = 0$ if x_{rs} is not in the basis; and $n_{N+ts} = 0$ if x_{N+ts} is not in the basis. There are $M + N - 1 + MP$ independent equations in the set **20, 21, 22** and $M + N + MP$ unknown n's (one to each basic variable and one to the non-basic variable x_{11}). It is therefore possible to solve this set of equations for the $(M + N - 1 + MP)$ n's associated with the basic variables in terms of n_{11} (associated with the non-basic variable to be introduced).

The values of the variables in the new solution are given by $x_{rs} + n_{rs}\theta$; $x_{N+ts} + n_{N+ts}\theta$. By selecting a suitable value of θ one of the variables may be reduced to zero while the others remain positive and a new basic feasible solution is obtained.

The value chosen is:

$$\theta = \operatorname*{Min}_{\substack{n_{rs}<0 \\ n_{N+ts}<0}} \left(-\frac{x_{rs}}{n_{rs}} ; -\frac{x_{N+ts}}{n_{N+ts}} \right)$$

In practice it is not necessary to write down all the equations **20, 21, 22.**

Change in Total Cost

To show that the method of altering the solution which has just been described is valid consider the introduction of x_{11}. Using equations **20, 21, 22** the change in total cost is

$$\sum_{i=1}^{N}\sum_{j=1}^{M} \theta n_{ij} c_{ij}$$

$$= \left\{ \sum_{i=1}^{N}\sum_{j=1}^{M} n_{ij}\left(r_j + s_i + \sum_{k=1}^{P} t_{jk}p_{ijk}\right) - n_{11}\left(r_1 + s_1 + \sum_{k=1}^{P} t_{1k}p_{11k}\right) + n_{11}c_{11}\right\}\theta$$

$$= \left\{ \sum_{j=1}^{M} r_j\left(\sum_{i=1}^{N} n_{ij}\right) + \sum_{i=1}^{N} s_i\left(\sum_{j=1}^{M} n_{ij}\right) + \sum_{k=1}^{P}\sum_{j=1}^{M} t_{jk}\left(\sum_{i=1}^{N} n_{ij}p_{ijk}\right) + n_{11}\left(c_{11} - r_1 - s_1 - \sum_{k=1}^{P} t_{1k}p_{11k}\right)\right\}\theta$$

$$= \left\{ \sum_{k=1}^{P}\sum_{j=1}^{M} t_{jk}\left(-n_{N+k-j}\right) + n_{11}\left(c_{11} - r_1 - s_1 - \sum_{k=1}^{P} t_{1k}p_{11k}\right)\right\}\theta$$

$$= n_{11}\,\theta\left(c_{11} - r_1 - s_1 - \sum_{k=1}^{P} t_{1k}p_{11k}\right),$$

since $t_{jk} = 0$ for those j, k for which x_{N+k-j} is in the basis and $n_{N+k-j} = 0$ for those j, k for which x_{N+k-j} is not in basis.

Consider now the introduction of the non-basic variable x_{N+11} and then the change in cost is:

$$\sum_{i=1}^{N}\sum_{j=1}^{M} \theta n_{ij} c_{ij} = \left\{ \sum_{j=1}^{M} r_j\left(\sum_{i=1}^{N} n_{ij}\right) + \sum_{i=1}^{N} s_i\left(\sum_{j=1}^{M} n_{ij}\right) + \sum_{j=1}^{M}\sum_{k=1}^{P} t_{jk}\left[\sum_{i=1}^{N} n_{ij}p_{ijk}\right]\right\}\theta$$

$$= \sum_{j=1}^{M}\sum_{k=1}^{P} t_{jk}\left(-n_{N+k-j}\right)\theta$$

$$= -\theta t_{11}n_{N11},$$

since $t_{jk} = 0$ for those j, k for which x_{N+k-j} is in the basis and $n_{N+k-j} = 0$ for those j, k for which x_{N+k-j} is not in basis (except for x_{N+11}).

A First Basic Feasible Solution

Using the technique of artificial variables it is shown that a first basic feasible solution can be found. From the usual arguments (e.g. Vajda, 1956) it can be shown that if the basis for the optimum solution contains any artificial variables (other than degenerate ones) then the problem does not have any solution.

Artificial Variables x_{0j} ($j = 1 \ldots M$); x_{i0} ($i = 1 \ldots N$) are introduced into equations **2** and **3** and extra slack variables x_{00}; x_{N+k-0} ($k = 1 \ldots P$) are also added. For consistency let $p_{00k} = 1$ ($k = 1 \ldots P$); $p_{0jk} = 0$ ($j = 1 \ldots M$; $k = 1 \ldots P$). The enlarged problem is:

Minimize $$\sum_{i=1}^{N}\sum_{j=0}^{M} x_{ij} c_{ij}$$

subject to $$\sum_{i=0}^{N} x_{ij} = b_j \; (j = 0 \ldots M),$$

$$\sum_{j=0}^{M} x_{ij} = a_i \quad (i = 1 \ldots N),$$

$$x_{N+k-j} + \sum_{i=0}^{N} x_{ij} p_{ijk} = d_{jk} \quad (j = 0 \ldots M; k = 1 \ldots P),$$

where all variables are non-negative and

$$b_0 = \sum_{i=1}^{N} a_i, \quad a_0 = \sum_{j=1}^{M} b_j, \quad d_{0k} = b_0 \quad (k = 1 \ldots P),$$

$$c_{00} = 0, \quad c_{0j} = M' \quad (j = 1 \ldots M), \quad c_{i0} = M' \quad (i = 1 \ldots N)$$

(M' is the usual high cost 'M-cost').

A basic feasible solution to this enlarged problem is: $x_{0j} = b_j$ ($j = 1 \ldots M$); $x_{i0} = a_i$ ($i = 1 \ldots N$); $x_{00} = \varepsilon$ (degenerate) $x_{+Nk-j} = d_{jk}$ ($j = 0 \ldots M$; $k = 1 \ldots P$).

Suppose that this solution is to be altered by the introduction of the non-basic variable x_{rs} (i.e. adding $n_{rs}\,\theta$ to it) then the following changes are made to the other variables:

$$x_{0s} - n_{rs}\,\theta;\ x_{r0} - n_{rs}\,\theta;\ x_{00} + n_{rs}\,\theta;$$
$$x_{N+ks} - (n_{rs}\,\theta)/p_{rsk}\,;\ x_{N+k} - n_{rs}\,\theta \quad (k = 1 \ldots P).$$

Numerical Example

A company making iron has a different type of furnace in each of three works. The works must receive a fixed weight of ore which is available in three different grades. For technical reasons the processing time of ore depends on its grade and the works to which it is sent. The furnaces are only available for a certain time in each week. The problem is to find the allocation which minimizes the production cost whilst satisfying the extra requirement that the amount of phosphorus is less than a certain critical level. The following information is available:

Cost, c_{ij}, of transport, purchase and processing ore per ton

		Works, j			*Tons available*	*Phos.*
		1	*2*	*3*	a_i	*content* p_i
	1	4	2	9	7	0·4
Ore, *i*	2	6	1	5	12	0·8
	3	7	8	3	6	0·7
Tons required	b_j	5	10	10		
Max. phos.	L_j	0·7	0·7	0·7		

Process time f_{ij} in hrs

		Works, j		
		1	*2*	*3*
	1	2	4	7
Ore, *i*	2	3	5	1
	3	9	8	6
Total time available	T_j	30	60	40

Let x_{ij} be the tonnage sent from *i* to *j*. Then it is required to

Minimize
$$\sum_{i=1}^{3}\sum_{j=1}^{3} x_{ij} c_{ij}$$

subject to
$$\sum_{i=1}^{3} x_{ij} = b_j$$

$$\sum_{j=1}^{3} x_{ij} = a_i$$

$$\sum_{i=1}^{3} p_i x_{ij} \leqslant L_j b_j$$

$$\sum_{i=1}^{3} f_{ij} x_{ij} \leqslant T_j \quad (x_{ij} \geqslant 0).$$

Computation

Table 1 shows the form of table that will be used for the calculations with the numbers replaced by the corresponding letters. For the numerical problem a first feasible solution can be found by inspection (see Table 2). Using this solution the associated fictitious costs are found.

Then $t_{11}, t_{12}, t_{22}, t_{32}$ are all zero since $x_{41}, x_{51}, x_{42}, x_{52}$ are in the basis. An arbitrary value of zero is assigned to r_1. Since x_{11} is in the basis, s_1 can be found, viz.

$$s_1 + 4t_{11} + 2t_{12} + r_1 = c_{11}, \quad \text{hence } s_1 + 0 = 4.$$

Similarly, x_{31} is in the basis so

$$s_3 = c_{31} - 7t_{11} - 9t_{12} - r_1 = 7 - 0 = 7;$$

x_{12} and x_{32} are both in the basis and so

$$s_1 + 4t_{21} + 4t_{22} + r_2 = c_{12}, \quad \text{whence } 4t_{22} + r_2 = -2$$
$$s_3 + 7t_{21} + 8t_{22} + r_2 = c_{32}, \quad \text{whence } 8t_{22} + r_2 = 1$$

and leading to $t_{22} = \frac{3}{4}$; $r_2 = -5$;
x_{22} is in the basis

$$s_2 + 8t_{21} + 5t_{22} + r_2 = c_{22} \quad \text{so } s_2 = 1 - 5(\tfrac{3}{4}) + 5 = 2\tfrac{1}{4}.$$

Finally, x_{13} and x_{23} are in the basis so

$$s_1 + 4t_{31} + 7t_{32} + r_3 = c_{13} \quad \text{so } 4t_{31} + r_3 = 5,$$
$$s_2 + 8t_{31} + t_{32} + r_3 = c_{23} \quad \text{so } 8t_{31} + r_3 = 2\tfrac{3}{4},$$

whence $t_{31} = -\frac{9}{16}$; $r_3 = 7\frac{1}{4}$.

All the fictitious costs have now been found and they are inserted in the correct place on Table 3. If q_{ij} is the difference between the true and fictitious costs then

$$q_{21} = 2\tfrac{1}{4} + 0 + 0 - 6 = -3\tfrac{3}{4}; q_{43} = -\tfrac{9}{16},$$

$$q_{33} = 7 + 7\tfrac{1}{4} + 7(-\tfrac{9}{16}) - 3 = 7\tfrac{5}{16}\ q_{52} = \tfrac{3}{4}.$$

The largest positive q is q_{33}. n_{33} is added to this variable and n_{rs} is added to all the basic variables x_{rs}. The n's satisfy the equations **20, 21, 22** and can be solved yielding:

For $j = 3$

$$\left.\begin{aligned} n_{13} + n_{23} + n_{33} &= 0 \\ 4n_{13} + 8n_{23} + 7n_{33} &= 0 \end{aligned}\right\} \text{so} \quad \begin{aligned} n_{13} &= -\tfrac{1}{4} n_{33} \\ n_{23} &= -\tfrac{3}{4} n_{33}, \end{aligned}$$

For $i = 2$

$$n_{22} + n_{23} = 0 \text{ so } n_{22} = \tfrac{3}{4} n_{33}$$
$$n_{53} = -6n_{33} - n_{23} - 7n_{13} = -3\tfrac{1}{2} n_{33}$$

For $j = 2$

$$\left.\begin{aligned} n_{12} + n_{22} + n_{32} &= 0 \\ 4n_{12} + 5n_{22} + 8n_{32} &= 0 \end{aligned}\right\} \text{so} \quad \begin{aligned} n_{32} &= -\tfrac{3}{16} n_{33} \\ n_{12} &= -\tfrac{9}{16} n_{33} \end{aligned}$$

For $i = 3$ $\quad n_{31} + n_{32} + n_{33} = 0 \quad \text{so } n_{31} = -\frac{13}{16} n_{33}$

For $i = 1$ $\quad n_{11} + n_{12} + n_{13} = 0 \quad \text{so } n_{11} = +\frac{13}{16} n_{33}$

$$n_{41} = -4n_{11} - 7n_{31} = +\tfrac{39}{16} n_{33}$$
$$n_{51} = -2n_{11} - 9n_{31} = +\tfrac{91}{16} n_{33}$$
$$n_{42} = -4n_{12} - 8n_{22} - 7n_{32} = -\tfrac{39}{16} n_{33}$$

Hence $n_{33} = \text{Min}\ (7\frac{1}{2} \times \frac{4}{3}, 2\frac{1}{4} \times 4, 15 \times \frac{2}{7}, 3\frac{7}{8} \times \frac{16}{3}, 1\frac{5}{8} \times \frac{16}{9}, 2\frac{1}{8} \times \frac{16}{13}, \frac{3}{8} \times \frac{16}{39})$

$= \frac{2}{13}$.

Using this value of n_{33} Table 4 is obtained. Similarly, Tables 5, 6 and 7 are found. All the fictitious costs are less than the true costs in Table 7 and the optimum has been found.

Table 1 Cost $= \sum_{i=1}^{N} \sum_{j=1}^{M} x_{ij}c_{ij}$

		$j=1$			$j=2$			$j=3$			
		t_{11}	t_{12}	r_1	t_{21}	t_{22}	r_2	t_{31}	t_{32}	r_3	
$i=1$	s_1		x_{11}			x_{12}			x_{13}		a_1
		f_{11}	p_1	c_{11}	f_{12}	p_1	c_{12}	f_{13}	p_1	c_{13}	
$i=2$	s_2		x_{21}			x_{22}			x_{23}		a_2
		f_{21}	p_2	c_{21}	f_{22}	p_2	c_{22}	f_{23}	p_2	c_{23}	
$i=3$	s_3		x_{31}			x_{32}			x_{33}		a_3
		f_{31}	p_3	c_{31}	f_{32}	p_3	c_{32}	f_{33}	p_3	c_{33}	
			x_{41}			x_{42}			x_{43}		
		1			1			1			
			x_{51}			x_{52}			X_{53}		
			1			1			1		
		L_1b_1	T_1	b_1	L_2b_2	T_2	b_2	L_3b_3	T_3	b_3	

Table 2 Cost $= 125\frac{1}{8}$

	t_{11}	t_{12}	r_1	t_{21}	t_{22}	r_2	t_{31}	t_{32}	r_3	
	0	0	0	0	$\frac{3}{4}$	-5	$-\frac{9}{16}$	0	$\frac{29}{4}$	
$s_1 = 4$		$2\frac{7}{8}+n_{11}$			$1\frac{5}{8}+n_{12}$			$2\frac{1}{2}+n_{13}$		7
	4	2	4	4	4	2	4	7	9	
$s_2 = 2\frac{1}{4}$					$4\frac{1}{2}+n_{22}$			$7\frac{1}{2}+n_{23}$		12
	8	3	6	8	5	1	8	1	5	
$s_3 = 7$		$2\frac{1}{8}+n_{31}$			$3\frac{7}{8}+n_{23}$			$+n_{33}$		6
	7	9	7	7	8	8	7	6	3	
		$8\frac{5}{8}+n_{41}$			$\frac{3}{8}+n_{42}$					
	1			1			1			
		$5\frac{1}{8}+n_{51}$						$15+n_{53}$		
		1			1			1		
	35	30	5	70	60	10	70	40	10	

Table 3 Cost $= 125\frac{1}{8} - 7\frac{5}{16}n33$

	t_{11}	t_{12}	r_1	t_{21}	t_{22}	r_2	t_{31}	t_{32}	r_3	
	0	0	0	0	$\frac{3}{4}$	-5	$-\frac{9}{16}$	0	$\frac{29}{4}$	
$s_1 = 4$		$2\frac{7}{8}+\frac{13}{16}n_{33}$			$1\frac{5}{8}-\frac{9}{16}n_{33}$			$2\frac{1}{2}-\frac{1}{4}n_{33}$		7
	4	2	4	4	4	2	4	7	9	
$s_2 = 2\frac{1}{4}$					$4\frac{1}{2}+\frac{3}{4}n_{33}$			$7\frac{1}{2}-\frac{3}{4}n_{33}$		12
	8	3	6	8	5	1	8	1	5	
$s_3 = 7$		$2\frac{1}{8}-\frac{13}{16}n_{33}$			$3\frac{7}{8}-\frac{3}{16}n_{33}$			$+n_{33}$		6
	7	9	7	7	8	8	7	6	3	
		$8\frac{5}{8}+\frac{39}{16}n_{33}$			$\frac{3}{8}-\frac{39}{16}n_{33}$					
	1			1			1			
		$5\frac{1}{8}+\frac{91}{16}n_{33}$						$15-\frac{7}{2}n_{33}$		
		1			1			1		
	35	30	5	70	60	10	70	40	10	

Table 4 Cost $= 124 - 6_{n_{21}}$

	t_{11}	t_{12}	r_1	t_{21}	t_{22}	r_2	t_{31}	t_{32}	r_3	
	0	0	0	-3	3	-2	-3	0	17	
$s_1 = 4$		$3+\frac{1}{3}n_{21}$			$1\frac{7}{13}$			$2\frac{6}{13}-\frac{1}{3}n_{21}$		7
	4	2	4	4	4	2	4	7	2	
$s_2 = 12$		$+n_{21}$			$4\frac{8}{13}$			$7\frac{5}{13}-n_{21}$		12
	8	3	6	8	5	1	8	1	5	
$s_3 = 7$		$2-\frac{4}{3}n_{21}$			$3\frac{11}{13}$			$\frac{2}{13}+\frac{4}{3}n_{21}$		6
	7	9	7	7	8	8	7	6	3	
	9									
	1			1			1			
		$6+\frac{25}{3}n_{21}$						$14\frac{6}{13}-\frac{14}{13}n_{21}$		
		1			1			1		
	35	30	5	70	60	10	70	40	10	

Table 5 Cost = $115 - 3_{n52}$

	t_{11}	t_{12}	r_1	t_{21}	t_{22}	r_2	t_{31}	t_{32}	r_3	
	0	0	0	$-\frac{3}{2}$	3	-8	$-\frac{3}{2}$	0	11	
$s_1 = 4$		$3\frac{1}{2}$			$1\frac{7}{13}+\frac{1}{13}n_{52}$			$1\frac{25}{26}-\frac{1}{13}n_{52}$		7
	4	2	4	4	4	2	4	7	9	
$s_2 = 6$		$1\frac{1}{2}$			$4\frac{8}{13}+\frac{3}{13}n_{52}$			$5\frac{23}{26}-\frac{3}{13}n_{52}$		12
	8	3	6	8	5	1	8	1	5	
$s_3 = 2\frac{1}{2}$					$3\frac{11}{13}-\frac{4}{13}n_{52}$			$2\frac{2}{13}-\frac{4}{13}n_{52}$		6
	7	9	7	7	8	8	7	6	3	
		9								
	1			1			1			
		$18\frac{1}{2}$			$+n_{52}$			$7\frac{6}{13}-\frac{14}{13}n_{52}$		
		1			1			1		
	35	30	5	70	60	10	70	40	10	

Table 6 Cost = $94\frac{3}{14} - \frac{21}{4}n_{31}$

	t_{11}	t_{12}	r_1	t_{21}	t_{22}	r_2	t_{31}	t_{32}	r_3	
	0	0	0	$-\frac{3}{4}$	0	1	$-\frac{149}{28}$	$-\frac{39}{14}$	$47\frac{3}{14}$	
$s_1 = 4$		$3\frac{1}{2}-\frac{1}{4}n_{31}$			$2\frac{1}{14}+\frac{1}{4}n_{31}$			$1\frac{3}{7}$		7
	4	2	4	4	4	2	4	7	9	
$s_2 = 6$		$1\frac{1}{2}-\frac{3}{4}n_{31}$			$6\frac{3}{14}+\frac{3}{4}n_{31}$			$4\frac{2}{7}$		12
	8	3	6	8	5	1	8	1	5	
$s_3 = 12\frac{1}{4}$		$+n_{31}$			$1\frac{5}{7}-n_{31}$			$4\frac{2}{7}$		6
	7	9	7	7	8	8	7	6	3	
		9								
	1			1			1			
		$18\frac{1}{2}-\frac{25}{4}n_{31}$			$6\frac{13}{14}+\frac{13}{4}n_{31}$					
		1			1			1		
	35	30	5	70	60	10	70	40	10	

Table 7 Cost $= 85\frac{3}{14}$

	t_{11}	t_{12}	r_1	t_{21}	t_{22}	r_2	t_{31}	t_{32}	r_3	
	0	0	0	$-\frac{3}{4}$	0	1	$-\frac{24}{7}$	$-\frac{9}{7}$	$\frac{32}{7}$	
$s_1 = 4$		$3\frac{1}{14}$			$2\frac{1}{2}$			$1\frac{3}{7}$		7
	4	2	4	4	4	2	4	7	9	
$s_2 = 6$		$\frac{8}{14}$			$7\frac{1}{2}$			$4\frac{2}{7}$		12
	8	3	6	8	5	1	8	1	5	
$s_3 = 7$		$1\frac{5}{7}$						$4\frac{2}{7}$		6
	7	9	7	7	8	8	7	6	3	
		9								
	1			1			1			
		$7\frac{11}{14}$			$12\frac{1}{2}$					
		1			1			1		
	35	30	5	70	60	10	70	40	10	

Conclusion

The iterative procedure described in this paper is not as difficult to operate as the equations suggest since many of the equations for fictitious costs and values of n can be found by inspection. The method takes advantage of the fact that the problem has a special form and so does not require the use of the simplex method.

References

GASS, S. I. (1958), *Linear Programming*, McGraw-Hill.

HALEY, K. B. (1963), 'The multi-index problem', *Operations Research,* vol. 2, no. 3.

VAJDA, S. (1956), *Theory of Games and Linear Programming*, Methuen.

VAJDA, S. (1962), *Readings in Mathematical Programming*, Pitman.

WILLIAMS, K. B., and HALEY, K. B. (1959), 'A practical application of linear programming in the mining industry', *Operational Research Q.*, vol. 10, no. 3.

17 E. M. L. Beale

Survey of Integer Programming

E. M. L. Beale, 'Survey of integer programming', *Operational Research Quarterly*, vol. 16, 1965, pp. 219–28.

Introduction

This paper surveys the progress that has been made with Integer Programming. This is the problem of solving linear, or possibly non-linear, programming problems when variables must take integer values. The field is divided into 'pure integer programming', when all the variables must be integers, and 'mixed integer programming', when only certain specified variables must be integers. The subject is also known as 'discrete programming', or 'integer linear programming' abbreviated to 'ILP', but I think that the term integer programming is preferred by the majority of workers in the field.

A particularly important type of integer programming problem is one in which the integer variables have to be either 0 or 1, depending on whether or not some action is taken. And pure integer programming has been successfully applied to a number of combinatorial problems of this type. More generally, the problem of finding a global optimum of some function over a non-convex region can be represented as a mixed integer programming problem. This important fact was recognized by Markowitz and Manne (1957). In these circumstances it does not seem important to distinguish between integer linear programming and integer non-linear programming. There is, of course, scope for studies of particular types of integer non-linear programming problems. Some pioneering work has been done in this area by Kelley (1960), by Künzi and Oettli (1963), and by Witzgall (1963).

I do not propose to say any more about applications of integer programming. The field was comprehensively surveyed by Dantzig (1960), and also in his book (1963).

From a practical point of view, the history of the subject is, very briefly, as follows:

In 1958 Gomory devised a method, known as the Method of Integer Forms, for solving pure integer programming problems. An outline of this was published at the time. For a long time the main reference was a Princeton University report, but this was eventually published (Gomory, 1963a). In 1960, he devised another method, the All-Integer Method (Gomory, 1963b). Several computer codes using variants of one or both methods have been written; and have successfully solved many real problems. However, their performance has not been nearly as predictable as that of ordinary linear programming codes – which are themselves rather unpredictable as regards running time. The most spectacular work in this area has been that of Glenn Martin. His code for the I.B.M. 7094 computer uses a variant of the Method of Integer Forms called the Accelerated Euclidean Algorithm (Martin, 1963). It has solved a number of problems with about 100 equations and 2000 variables. Up to the beginning of 1964 the largest single problem had about 215 equations and about 2600 variables.

The field of mixed integer programming is less far advanced. Gomory has extended his method of integer forms to deal with continuous as well as integer variables (Gomory, 1960). Various computer codes have been written using this method. Reports on their performance have varied from good to disappointing, but I am not aware that they have solved any large problems. William White has shown that, unless the objective function itself must be integral, the method may not terminate.

A mixed integer programming procedure, published by Healy (1964) under the title *Multiple Choice Programming* has been programmed for the I.B.M. 7090 computer and found to solve practical problems, although its theoretical status is obscure. Its emphasis on problems in which a sum of non-negative integer-valued variables must add up to 1 is appropriate, since many practical problems have this structure.

Driebeck (1964) and Dakin (1964) have developed programmes using a 'branch and bound' method for mixed integer programming.

Algorithms

The rest of this paper will be devoted to algorithms for both pure and mixed integer programming. I have classified these algorithms according to the type of approach taken. Since many people, including me, thought until 1958 that a general method of solving integer programming problems was self-evidently impossible, it is remarkable that there are now four distinct approaches capable of solving real problems in this area effectively. These can be described briefly as:

1. Cutting-plane methods.
2. Primal methods.
3. Branch and bound methods.
4. Partial enumeration methods.

Since pure integer programming seems to be easier than mixed integer programming, methods of reducing a mixed problem to a pure one are important. And in this context we should explicitly consider Benders's (1962) partitioning schemes for mixed variables problems.

I suspect that no single approach is suitable for all integer programming problems, so it is entirely appropriate that research is now proceeding on a broad front. This must be, and is being, supported by computational trials, since the mathematically most elegant method of solving a numerical problem may well not be the most efficient.

A few general remarks are needed before discussing some particular algorithms. First a piece of terminology. I hope that we can agree to restrict the term 'feasible solution' to one that is really feasible, i.e. one where all variables that have to be integral are integral, in addition to all variables that have to be non-negative being non-negative. The term 'non-negative' can be used to describe a solution in non-negative variables that may contain illegal fractions.

Next I want to consider the objective function. The theory of all Gomory's methods is expressed more neatly in terms of an objective function to be maximized, so I shall adopt this convention throughout this paper. Furthermore, if we are maximizing some function of n variables $x_1, x_2, \ldots, x_n$, it is often convenient to think of the objective function as an additional variable x_0. So the problem is to maximize x_0 over some set of points, 'the feasible set', in $(n + 1)$-dimensional space.

So far, cutting-plane methods for integer programming have been easily the most successful. They are based on the fact that the point where x_0 is maximized over the feasible set is also the point where x_0 is maximized over the convex hull of the feasible set. Furthermore, in either a pure or a mixed integer linear programming problem the convex hull is defined by a finite number of linear inequalities. The problem is to discover these inequalities, or at least the relevant ones. The cutting-plane approach is to start with some region that certainly includes the convex hull of the feasible set. If one finds a point that maximizes x_0 in this region that is not feasible, then one generates an additional linear inequality constraint, or cutting plane. This constraint will be one that is not satisfied at the current trial solution but that must be satisfied at all feasible solutions. This approach was used in an unsystematic way by Dantzig, Fulkerson and Johnson (1954), and by Markowitz and Manne (1957). But the subject of integer programming really began when Gomory showed how one could apply this idea in an automatic and systematic way that guarantees convergence in a finite number of steps. Since the problem has a trivial finite solution if one enumerates all possible sets of values of the integer variables, it is curious that it should be so difficult to guarantee convergence for more high-powered methods. But this is the case. It is based on the massive 'degeneracy' (in linear programming terminology) that often arises in integer programming problems and that cannot be removed by small perturbations. I shall not attempt to give the convergence proofs here, but I will explain the derivation of the cuts in Gomory's three published methods. Incidentally I should like to mention here for the benefit of those who read French more easily than English that these methods are all described in a book by Simmonard (1962).

In two of the methods we assume that all the coefficients in the original definition of the problem are integers. (This is to avoid numerical difficulties.) And in practice it is important that they should be fairly small integers. The exception is the All-

Integer Method, which may start with inequalities containing arbitrary coefficients.

In the original method for pure integer programming, the Method of Integer Forms, one starts by solving the problem ignoring the requirement that the variables must be integers. If this happens to produce a solution in integers, then the problem is solved. Otherwise one considers the expression for some basic fractional variable in terms of the non-basic variables. Let this be:

$$x_s = a_{s0} - \sum_{j=1}^{n} a_{sj}\, t_j$$

where t_j denotes the jth non-basic variable.

We now write the coefficients a_{sj} in the form:

$$a_{sj} = n_{sj} + f_{sj},$$

where the n_{sj} are all integers, though not necessarily positive integers, and the f_{sj} are all fractions such that:

$$0 \leqslant f_{sj} < 1.$$

Then we consider the expression:

$$s = \; - f_{s0} + \sum_{j=1}^{n} f_{sj}\, t_j.$$

This expression must be an integer, since it differs from $- x_s$ by an integer. On the other hand, it cannot be less than $- f_{s0}$ since the expression $\sum f_{sj} t_j$ must be non-negative. And the smallest integer that is not less than $- f_{s0}$ is zero. So s must be a non-negative integer, and we can introduce it as a new variable of our problem. Its value at the current trial solution is $- f_{s0}$. So the constraint that $s \geqslant 0$ cuts off this solution and enables us to continue optimizing using the dual simplex method.

One can generate such a cut from the expression for any basic variable that is fractional at the current trial solution, or more generally from any integer combination of such variables. But, in fact, Gomory has shown that there are only $D - 1$ different cuts obtainable in this way, where D is the modulus of the determinant of the current basis. He has also shown that, given certain rules about which cut to take, the procedure must terminate at the optimum solution if there is one. (If there is no feasible solution to the problem, the procedure will prove in a finite number of steps that there is no solution in which the objective function exceeds any prescribed lower bound.)

Martin's procedure is a variant of this method related to a suggestion made in Gomory's paper and attributed to me. This is that one should first find a solution that is non-negative but possibly fractional, then suspend the requirement that the basic variables must be non-negative while adding cuts to make the system integral, then restore non-negativity, and so on. Obviously in these circumstances one will choose each cut so as to obtain a small pivot, since at each iteration D is multiplied by the magnitude of the pivot. But Martin's procedure differs from any envisaged by either Gomory or me in that it does not always maintain dual feasibility.

Martin's success in solving large problems is due partly to his procedure for choosing cuts, but partly to the fact that these particular problems are formulated as directed network problems with some extra constraints on the flows through particular arcs. These problems tend to be easy ones, since without the extra constraints they naturally have integer solutions. The approach is described by Martin (1964).

Gomory's second method for pure integer programming, the All-Integer Method, proceeds rather differently. One does not start by solving the problem in continuous variables, since the method involves keeping the coefficients integral throughout. One starts by making the problem dual feasible, perhaps by adding an artificial constraint that the sum of the non-basic variables is less than or equal to some arbitrarily large number. After this every single pivotal row is a new cut generated in such a way as to make the pivot equal to -1. This ensures that an integral tableau remains integral.

We now consider how to generate a pivotal row with this valuable property.

Suppose that x_s is some variable that is negative in the current trial solution, and let the expression for x_s in terms of the non-basic variables be:

$$x_s = a_{s0} - \sum_{j=1}^{n} a_{sj}\, t_j,$$

where a_{s0} is negative.

Now let λ be any positive quantity, and write:

$$a_{sj} = \lambda n_{sj} + f_{sj},$$

where n_{sj} is an integer and $0 \leqslant f_{sj} < \lambda$.

Consider the expression:

$$s = n_{s0} - \sum_{j=1}^{n} n_{sj}\, t_j.$$

This expression must obviously be integral. Furthermore it must be positive, since:

$$f_{s0} - \sum_{j=1}^{n} f_{sj}\, t_j \leqslant f_{s0} < \lambda$$

so that if $s \leqslant -1$ we would have:

$$x_s = \lambda s + f_{s0} - \sum_{j=1}^{n} f_{sj}\, t_j < -\lambda + \lambda = 0,$$

which is contrary to hypothesis.

We note that for any value of λ, $n_{s0} \leqslant -1$ if $a_{s0} < 0$. So the new constraint that $s \geqslant 0$ certainly cuts off the current trial solution. Furthermore, we can make the pivot equal to -1 by taking λ large enough. This is because for arbitrarily large λ all $n_{sj} = 0$ or -1 depending on whether a_{sj} is non-negative or not. In fact, given any expression from which to derive the cut, it is a moderately straightforward matter to derive the smallest value of λ that makes the pivot equal to -1. And this is the most powerful cut of the family.

Just as constraints for the method of Integer Forms can be generated from any linear combination of rows provided it is integral, constraints for the All-Integer Method can be generated from any linear combination of rows provided it is non-negative. The most interesting development in this direction is some work of Gomory's in which one solves a linear programming sub-problem at each iteration to choose a linear combination of rows that puts the pivot in the largest possible lexicographic column. Aesthetically this is a satisfactory procedure, since it finds the best possible cut in a clearly defined sense. Balinski (1964) reports a private communication from Dr Gomory saying that this procedure has a very good effect on the number of iterations, but that the computer programming for the solution of the subproblems needs to be streamlined to make the procedure really attractive.

Gomory's procedure for the mixed integer problem also follows the cutting-plane approach. It is related to the Method of Integer Forms. One first solves the problem ignoring the requirement that any variables must be integral. If some variable that should be an integer, say x_s, equals $n_{s0} + f_{s0}$, where n_{s0} is a non-negative integer and $0 < f_{s0} < 1$, then one considers the expression for x_s in terms of the current non-basic variables. This can be written:

$$x_s = n_{s0} + f_{s0} - \Sigma^{+} a_{sj}\, t_j + \Sigma^{-} b_{sj}\, t_j,$$

where the non-basic variables t_j are divided into two groups such that all $a_{sj} \geqslant 0$ and all $b_{sj} \geqslant 0$.

Now consider the expression:

$$s = -f_{s0} + \Sigma^{+} a_{sj}\, t_j + \left(\frac{f_{s0}}{1-f_{s0}}\right) \Sigma^{-} b_{sj}\, t_j.$$

This expression is negative at the current trial solution. Yet it must be non-negative for any integral value of x_s. This is because if $x_s - n_{s0} \leqslant 0$ we must have:

$$s \geqslant -f_{s0} + \Sigma^{+} a_{sj}\, t_j - \Sigma^{-} b_{sj}\, t_j = -(x_s - n_{s0}) \geqslant 0,$$

whereas if $x_s - n_{s0} \geqslant 1$ we must have:

$$\left(\frac{1-f_{s0}}{f_{s0}}\right) s \geqslant -(1-f_{s0}) - \Sigma^{+} a_{sj}\, t_j + \Sigma^{-} b_{sj}\, t_j$$

$$= x_s - n_{s0} - 1 \geqslant 0.$$

Consequently $s \geqslant 0$ in all circumstances, and this defines a valid cut. If any of the non-basic variables must be integral then we can sharpen the cut by working with the expression for x_s plus an integer combination of non-basic integer-valued variables, chosen so that $a_{sj} \leqslant f_{s0}$, or alternatively $b_{sj} \leqslant 1 - f_{s0}$, for all such variables. But note that s is not necessarily integral.

So much for the cutting-plane approach. We now turn to the primal approach. This involves progressing from one feasible solution to another with a larger value of x_0, or else proving that no better solution can exist. I would exclude methods of finding local optima (in some sense) that may not be global optima, on the grounds that they are not truly integer programming methods at all – though they may well provide the best available approach to many real problems. This is particularly true if one also obtains an upper bound on the possible value of the objective function by starting towards a dual solution using a cutting-plane method.

Primal integer programming may ultimately prove the most powerful approach to the problem. What one would like to do is to proceed from one vertex to a better neighbouring vertex of the convex hull of the feasible set. One can go some way towards this by generating pivotal rows using the λ-transformations of the All-Integer Method. The difficulty is that in this case a_{s0} must be non-negative – since the current trial solution must be feasible. And if $a_{s0} < a_{sj}$, where t_j is the variable to be removed from the set of non-basic variables, then the derived pivotal row has a zero constant term, and no progress is possible. The problem was discussed by Ben-Israel and Charnes (1962). Nevertheless Gomory was able to announce a finite method for pure integer programming on these lines at the Chicago Symposium on Mathematical Programming in 1962. This has not proved computationally successful. Harris (1964) has devised a method for the mixed integer programming problem on these lines, using Benders's (1962) partitioning scheme. Her method produces the best possible new non-basic variable in a certain sense, and therefore looks very promising.

Our third approach is the Branch and Bound technique. This involves setting up a tree of linear programming problems. The root of the tree is the problem in which all integrality constraints are ignored. If any integer variable is fractional at the optimal solution to this problem, then one considers one such variable, say $x_s = n_{s0} + f_{s0}$, and sets up two alternative problems. One contains the additional constraint $x_s \leqslant n_{s0}$. The other contains the additional constraint $x_s \geqslant n_{s0} + 1$. These constraints can be thought of as branches of the tree. One can then form sub-branches on the values of other variables that must be integers. The problem is solved when one has a feasible optimal solution on one branch of the tree, and lower values of x_0 on all other branches, whether feasible or not.

Such an approach was proposed by Land and Doig (1960). Streamlined versions have been developed by Driebeck (1964) and Dakin (1964). Although this approach will not produce spectacular results on large problems that yield easily to cutting-plane methods, I think it may perform very serviceably on mixed problems with only a moderate number of integer variables. I am greatly encouraged in this belief by the success reported by Little *et al.* (1963) in solving travelling salesman problems in this way.

This work also throws light on the important problem of choosing which variable to branch on. It appears that one should choose that variable for which the worse of the two alternative inequalities is as bad as possible. This means that there is an excellent chance that one will only need to explore one of the two branches created.

The fourth approach, partial enumeration, may seem even less powerful. But it was used successfully on small problems by Benders, Catchpole and Kuiken (1959), and it has been developed further by Balas (1964a and b).

I want to finish with a few remarks about Benders's (1962) partitioning procedure. This is usually presented in terms of duality, which I personally find confusing. The principle can be explained in an elementary way as follows:

Suppose that we can express the objective function x_0 in the form:

$$x_0 = a_{00} - \sum_{j=1}^{n} a_{0j} t_j - \sum_{k=1}^{n} b_{0k} y_k,$$

where the b_{0k} are non-negative and the variables y_k are constrained to be non-negative. Then we have the inequality:

$$x_0 \leqslant a_{00} - \sum_{j=1}^{n} a_{0j} t_j.$$

One situation in which this fact is useful is where the y_k are continuous variables and the t_j are integer variables. The expression for x_0 will be in the required form if one maximizes it for fixed values of the t_j. The question of what fixed values to use is a separate one. But if one chooses those values that would maximize x_0 if the inequalities derived so far were the only operative ones, then one obtains a procedure that is dual to the well-known Dantzig–Wolfe (1960) Decomposition procedure.

It is perhaps worth mentioning that these inequalities were used in a primitive way in an unpublished Princeton report of mine on mixed integer programming (1958). But I did not then appreciate that x_0 could be treated as a variable. My inequalities used the current numerical value of x_0. From a formal point of view this simplified the situation, since it meant that my 'master problem', to use decomposition terminology, was a pure integer programming problem instead of having one continuous variable, namely x_0. But in practice it is obviously very inefficient not to sharpen the inequality when one obtains a better value of x_0.

Conclusions

Integer programming is a fast-moving field, and I cannot pretend to be aware of all that is happening in it. But I hope that the paper explains how progress is now being made. This does not mean that all problems that can be formulated as integer programming problems can be solved as such. It seems unlikely that it will ever be easier to solve an integer programming problem than a straightforward linear programming problem of the same size; and some integer programming formulations are vast problems even if one ignores the requirement that the variables must be integers. But we can now solve some large integer programming problems, and we shall soon be able to solve many more

References

BALAS, E. (1964a), 'Un algorithm additif pour la résolution des programmes linéaires en variables bivalentes', *C.R. Acad. Sci. Paris*, vol. 258, pp. 3817–20.

BALAS, E. (1964b), 'Extension de l'algorithme additif à la programmation en nombres entiers et à la programmation non linéaire', *C.R. Acad. Sci. Paris*, vol. 258, pp. 5136–9.

BALINSKI, M. L. (1964), *On Finding Integer Solutions to Linear Programs*, *Mathematica*, Princeton.

BEALE, E. M. L. (1958), 'A method of solving linear programming problems when some but not all of the variables must take integral values', Statistical Techniques Research Group, *Tech. Rep.*, no. 19, Princeton University.

BENDERS, J. F. (1962), 'Partitioning procedures for solving mixed-variables programming problems', *Numerische Mathematik*, vol. 4, pp. 238–52.

BENDERS, J. F., CATCHPOLE, A. R., and KUIKEN, C. (1959), 'Discrete variables optimization problems', paper presented at the RAND Symposium on Mathematical Programming, March.

BEN-ISRAEL, A., and CHARNES, A. (1962), 'On some problems of diophantine programming', *Cahiers Centre Etudes Rech. Oper.*, vol. 4, pp. 215–71.

DAKIN, R. J. (1964), 'A mixed-integer programming algorithm', Basser Computing Laboratory, School of Physics, University of Sydney, *Tech. Rep.*, no. 31.

DANTZIG, G. B. (1960), 'On the significance of solving linear programming problems with some integer variables', *Econometrica*, vol. 23, pp. 30–44.

DANTZIG, G. B. (1963), *Linear Programming and Extensions*, Princeton University Press.

DANTZIG, G. B., FULKERSON, D. R., and JOHNSON, S. M. (1954), 'Solution of a large-scale travelling salesman problem', *J. Operat. Res. Soc. Amer.*, vol. 2, pp. 393–410.

DANTZIG, G. B., and WOLFE, P. (1960), 'Decomposition principle for linear programs', *Operat. Res.*, vol. 8, pp. 101–11.

DRIEBECK, N. J. (1964), 'A method for the solution of mixed integer and non-linear programming problems by linear programming methods', paper presented at the International Symposium on Mathematical Programming, London School of Economics, July 1964.

GOMORY, R. E. (1960), 'An algorithm for the mixed integer problem', RM-2597 RAND Corporation.

GOMORY, R. E. (1963a), 'An algorithm for integer solutions to linear programs', in R. L. Graves and P. Wolfe (eds.), *Recent Advances in Mathematical Programming*, McGraw-Hill, pp. 269–302. First issued as Princeton-I.B.M. Mathematics Research Project Technical Report no. 1, 1958.

GOMORY, R. E. (1963b), 'All-integer integer programming algorithm', in J. F. Muth and G. L. Thompson (eds.), *Industrial Scheduling*, Prentice Hall, pp. 193–206. First issed as I.B.M. Research Center Research Report RC-189, 1960.

HARRIS, P. M. J. (1964), 'An algorithm for solving mixed integer linear programmes', *Operational Research Q.*, vol. 15, pp. 117–32.

HEALY, JR, W. C. (1964), 'Multiple choice programming', *Operat. Res.*, vol. 12, pp. 122–38.

KELLEY, JR, J. E. (1960), 'The cutting-plane method for solving convex programs', *J. Soc. Industr. Appl. Math.*, vol. 8, pp. 703–12.

KUNZI, H. P., and OETTLI, W. (1963), 'Integer quadratic programming', in R. L. Graves and P. Wolfe (eds.), *Recent Advances in Mathematical Programming*, McGraw-Hill, pp. 303–8.

LAND, A. H., and DOIG, A. G. (1960), 'An automatic method of solving discrete programming problems', *Econometrica*, vol. 28, pp. 497–520.

LITTLE, J. D. C., MURTY, K. G., SWEENEY, D. W., and KAREL, C. (1963), 'An algorithm for the traveling salesman problem', *Operat. Res.*, vol. 11, pp. 969–72.

MARKOWITZ, H. M., and MANNE, A. S. (1957), 'On the solution of discrete programming problems', *Econometrica*, vol. 25, pp. 84–110.

MARTIN, G. T. (1963), 'An accelerated euclidean algorithm for integer linear programming', in R. L. Graves and P. Wolfe (eds.), *Recent Advances in Mathematical Programming*, McGraw-Hill, pp. 311–18.

MARTIN, G. T. (1964), 'Directed network models for discrete programming', paper presented at the International Symposium on Mathematical Programming, London School of Economics, July.

SIMMONARD, M. (1962), *Programmation Linéaire*, Dunod.

WITZGALL, C. (1963), 'An all-integer programming algorithm with parabolic constraints', *J. Soc. Industr. Appl. Math.*, vol. 11 pp. 855–71.

18 N. Agin

Optimum Seeking with Branch and Bound

N. Agin, 'Optimum seeking with branch and bound', *Management Science*, vol. 13, 1966, pp. B176–85.

Introduction

In the recent past, branch and bound algorithms have been formulated for solving a wide variety of combinatorial problems. Some references to these are given at the end of the paper. We provide here an expository description of the use of branch and bound for optimum seeking in general. An objective is to convey to the unfamiliar reader the basic philosophy behind the technique. Also, we wish to provide the notation and definitions necessary in order for the reader to become aware of the applicability of branch and bound as a general approach to optimum seeking. Such a generalization should provide a basis for the recognition of problems suitable for solution by branch and bound as well as provide a framework for the development and description of efficient solution algorithms. Similar objectives are to be found in the work of Golomb and Baumert (1965).

We begin by defining what we mean by a combinatorial problem. Two such problems – that of the travelling salesman and job shop scheduling are provided as illustrations. We next develop the definitions and notation necessary to generalize a solution approach to such problems. Given these, we are able to define a branch and bound algorithm for obtaining the solution to any combinatorial problem. The branch and bound algorithms which presently exist for solving the travelling salesman problem and the three machine job shop scheduling problem are used as specific examples. Two aspects of the algorithm are then discussed. The first assures that the branch and bound algorithm will solve the combinatorial problem and the second that it may be possible to do so in less than complete enumeration. Finally the characteristics of a branch and bound algorithm which influence its computational efficiency are isolated and briefly discussed. Illustrations of these characteristics from the already mentioned example problems are given.

Combinatorial Problems

A combinatorial problem is here defined to be one of assigning discrete numerical values to some finite set of variables X, in such a way as to satisfy a set of constraints and minimize some objective function, $Z(X)$. A large range of flexibility is permitted in both the constraint set and the objective function. The objective function, $Z(X)$, is required only to be uniquely determined by a given set of values for X. Several applications of branch and bound exist in which $Z(X)$ is defined to take on values of 0 or ∞; indicating only that a solution is feasible or not. The ability to solve combinatorial problems with non-linear, discontinuous, or even non-mathematically defined objective functions must be recognized as an important advantage of branch and bound methods. Similar flexibility exists in the constraints.

We have selected two examples of combinatorial problems. Although they are not representative of the range of possible combinatorial problems they are indicative of the types of problems for which branch and bound methods are applicable. Algorithms for the solution of both have already been developed so that having introduced the problems, we will be able to use the branch and bound algorithms for solving these problems as later illustrations.

(1) The travelling salesman problem: Assign values of 0 or 1 to variables $X = \{x_{ij}\}$ where x_{ij} is one if the salesman travels from city i to city j and 0 otherwise. The constraints of the problem are such that the salesman must start at some city; visit each of the other cities once and only once; and return to his original city. If, associated with travelling from city i to city j, there is some cost c_{ij} then the objective function is to minimize

$$Z(X) = \sum_i \sum_j c_{ij}\, x_{ij}$$

which represents the total cost of visiting each city.

(2) The three-machine job shop scheduling problem: Assign

integer values to variables $X = \{x_{ij}\}$ where x_{ij} is the start time of job i on machine j; $j = 1, 2, 3$. The constraints of the problem are such that a job cannot be processed on machine 3 before it has been completed on machine 2 and it cannot be processed on machine 2 before it has been completed on machine 1. Given the processing time, t_{ij}, of job i on machine j the problem is to sequence the jobs on each machine so that the total time for the completion of all jobs is a minimum. That is, to minimize the objective function

$$Z(X) = \max_i (x_{i3} + t_{i3}).$$

Since $x_{i3} + t_{i3}$ is the time job i is completed on machine 3, the maximum of these numbers is the time at which the latest job is completed. It is that time which is to be minimized.

Definitions

In order to describe and define a branch and bound algorithm some definitions and notation are needed. The constraints of the combinatorial problem shall be designated as being either implicit or explicit. This dichotomy will be useful in developing the definitions and later in describing several branch and bound algorithms. The definitions are:

Implicit constraints: Constraints which will be satisfied by the manner in which the branch and bound algorithm is constructed, e.g., that the variables take on integer values.

Explicit constraints: Constraints which will require procedures for recognition as an integral part of the branch and bound algorithm, e.g. constraint equations which the values of the variables must satisfy.

Solution: An assignment of numerical values to X that satisfies all the implicit constraints.

Feasible solution: An assignment of numerical values to X that satisfies all the constraints.

s: A particular assignment of numerical values to X which satisfies all the implicit constraints, i.e., s is some solution to the combinatorial problem.

Ω: $\Omega = \{s\}$ the set of all solutions to the combinatorial problem.

S: A subset of Ω, i.e. some collection of solutions to the combinatorial problem.

Partition of Ω*:* An exhaustive division of Ω into disjoint subsets $S_1, S_2, \ldots, S_n$. That is, a collection of subsets $S_1, S_2, \ldots, S_n$ with the properties

$$S_1 \cup S_2 \cup \ldots \cup S_n = \Omega,$$
$$S_i \cap S_j = \varphi, \qquad i \neq j.$$

Branching: The process of partitioning a subset S into m disjoint subsets $S_1, S_2, \ldots S_m$ i.e., where

$$S_1 \cup S_2 \cup \ldots \cup S_m = S,$$
$$S_i \cap S_j = \varphi, \qquad i \neq j.$$

A process of successively branching can be visualized as the creation of a tree. The initial node of the tree represents Ω, the set of all solutions. Branches are created by the process defined as branching and nodes of the tree represent the subsets S of Ω. With each node N there is associated some subset S.

The example tree in Figure 1 is used to illustrate these concepts for the case $m = 2$. Node $N = 1$ corresponds to Ω, the set of all solutions. Nodes 2 and 3 are associated with disjoint subsets S_1 and S_2 of Ω. No branching has yet taken place from nodes 3, 4 and 5. This relation, between subsets of Ω and branching to the creation of a tree, can be formalized with the following definitions.

S(N): the subset $S \subset \Omega$ represented by the node N.

Intermediate node: A node N from which no branching has yet taken place.

Final node: An intermediate node N for which $S(N)$ consists of a single solution, s.

L(N): A lower bound on the value of the objective function for all solutions associated with node N, i.e., $L(N)$ is some fixed number with the property that it does not exceed $Z(s)$ for each $s \in S(N)$.

A Branch and Bound Algorithm

Using these definitions it is possible to define a branch and bound algorithm as follows:

Branch and bound algorithm: A set of rules for (1) branching from nodes to new nodes, (2) determining lower bounds for the

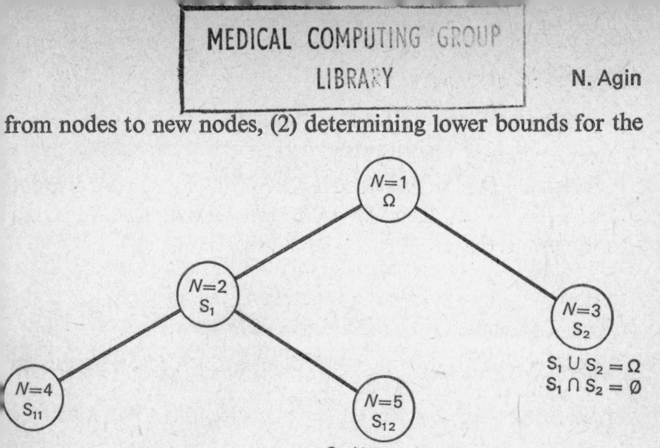

Figure 1

new nodes, (3) choosing an intermediate node from which to branch next, (4) recognizing when a node contains only infeasible or non-optimal solutions and (5) recognizing when a final node contains an optimal solution.

The branch and bound algorithm as defined above will solve the combinatorial problem. This can be referred to as its branching characteristic.

Branching characteristic: The branching characteristic of a branch and bound algorithm applicable to a given combinatorial problem guarantees that an optimal solution will eventually be obtained.

Since Ω is finite the process of branching will eventually yield any given solution, s, as a final node, unless stopped previously. That is, the branching process would lead to a partition of Ω for which each subset S_i obeys, by the definition of a partition,

$$S_1 \bigcup S_2 \bigcup \ldots \bigcup S_n = \Omega,$$
$$S_i \bigcap S_j = \varphi, \qquad i \neq j,$$

and $S_i = s_i$. Thus, all possible final nodes may eventually be generated and the optimal solution can be obtained.

The branching characteristic merely promises complete enumeration by the branch and bound algorithm unless it can find an optimal solution before doing so. That this may be the case is demonstrated by its bounding characteristic.

Bounding characteristic: The bounding characteristic of a branch and bound algorithm applicable to a given combinatorial problem furnishes the possibility of recognizing an optimal solution prior to complete enumeration.

Consider some final node N^* for which $S(N^*) = s^*$. If $Z(s^*) \leqslant L(N)$ for each intermediate node N then s^* is an optimal solution and there need be no further branching from the nodes N. This follows from the definition of $L(N)$, i.e.,

$$Z(s^*) \leqslant L(N) \leqslant \min_{s \varepsilon S(N)} Z(s) \leqslant Z(s) \quad \text{for each} \quad s \,\varepsilon\, S(N).$$

Verbally, for each $s \,\varepsilon\, S(N)$ there exists an unknown value for $Z(s)$. $L(N)$, however, is defined to be equal to or less than even the smallest of these $Z(s)$. Thus, if a solution has been found with value $Z(s^*)$ which is no greater than $L(N)$, for each N, then it must be optimal.

To illustrate how the preceding definitions can be used to define a specific branch and bound algorithm we describe the algorithms which have been developed for solving the travelling salesman and the three-machine job shop scheduling problems. The method for solving the travelling salesman problem is due to Little, *et al.* (1963). In this context, Ω, the class of all solutions to the combinatorial problem is equivalent to all the tours the travelling salesman can take. Subsets S of Ω represent partially determined tours, i.e. tours in which some transitions from city to city are expressly included, some which are excluded and others not yet determined. In the terms of the general branch and bound algorithm the Little algorithm can be described as follows:

(1) Branching from nodes to new nodes: Given the intermediate node N from which next to branch create two new nodes, $N + 1$ and $N + 2$. $S(N + 1)$ contains all tours for which there is a transition from city i to city j and $S(n + 2)$ contains all tours for which there is no transition from city i to city j. Cities i and j are selected so as to increase most $L(N + 2)$.

(2) Determining lower bounds for the new nodes: The lower bound for any node N is the sum of the cost elements for the city pairs already committed to $S(N)$ plus a lower bound on the costs of going to each of the cities not yet committed to $S(N)$. This lower bound is calculated by a row and column reduction on the cost matrix for the remaining cities.

(3) Choosing an intermediate node from which to next branch: Branching takes place from the intermediate node N for which $L(N)$ is a minimum.

(4) Recognizing when a node contains only infeasible or non-optimal solutions: Infeasible solutions will occur if some city is visited twice before all cities are visited, i.e., a sub-tour is formed. They can be recognized by putting appropriate infinities in the cost matrix after each branching. If, later, $L(N) = \infty$ for some node N then $S(N)$ contains only infeasible solutions. The constraint against sub-tours, in this case, is explicit. Non-optimal solutions are recognized by comparing $L(N)$ to the best $Z(s)$ to date.

(5) Recognizing when a final node contains an optimal solution: A final node containing s is recognized as optimal if $Z(s) \leqslant L(N)$ for each intermediate node N.

The branch and bound algorithm which we describe for the three-machine job shop scheduling problem is due to Ignall and Schrage (1965). A basically identical algorithm is also found in Lomnicki (1965). For the case of three machines it has been shown that in the optimal schedule the sequence of jobs on each machine is identical i.e., if job i_2 immediately follows job i_1 on machine 1, then i_2 will also do so on machines 2 and 3. Thus, for this problem Ω represents all the permutations of the integers $\{1, 2, \ldots, n\}$ one of which is the optimum sequence for scheduling the n jobs. Subsets S of Ω will represent the partial permutations of the integers $\{1, 2, \ldots, n\}$, i.e. permutations in which the first k integers are specified and the remaining $n - k$ are unspecified. The algorithm itself is described as follows:

(1) Branching from nodes to new nodes: A node N represents k jobs already sequenced and $n - k$ jobs remaining to be se-sequenced. Given an intermediate node N the branching rule is to create $n - k$ additional nodes where each represents one of the $n - k$ remaining jobs sequenced as $k + 1^{st}$.

(2) Determining lower bounds for the new nodes: The lower bound for any node N is the maximum of

(a) the time the kth job is completed on machine 3 plus the total time for the remaining jobs to be processed on machine 3.

(b) the time the kth job is completed on machine 2 plus the total time for the $n - k$ remaining jobs to be processed on machine 2 plus the time for the job of the remaining $n - k$ jobs which has the shortest processing time on machine 3.

(c) the time required for all jobs to be processed on machine 1 plus the times for the job of the remaining $n - k$ jobs which has the shortest processing times on machines 2 and 3.

(3) Choosing an intermediate node from which next to branch: Branching takes place from the intermediate node N for which $L(N)$ is a minimum.

(4) Recognizing when a node contains only infeasible or non-optimal solutions: All the constraints of the job shop scheduling problem are implicit so that all nodes generated contain feasible solutions. A node containing all non-optimal solutions will have $L(N)$ greater than the currently best $Z(s)$.

(5) Recognizing when a final node contains an optimal solution: A final node containing s is recognized as optimal if $Z(s) \leqslant L(N)$ for each intermediate node N.

Computational Considerations

A branch and bound algorithm has been defined in terms of a set of rules for branching, bounding, choosing intermediate nodes, recognizing infeasible or non-optimal nodes and determining optimal solutions. The first three of these are believed to be of primary importance in terms of their effect on computation time. We discuss here some heuristic considerations involved in formulating these rules.

The importance of the particular aspects we discuss have not been empirically verified nor do they form a complete set. They should, however, suggest to the reader some of the considerations involved in developing efficient branch and bound algorithms. This is important, since, in general, the time required to solve combinatorial problems by branch and bound increases rapidly with problem size. As experience is obtained with these algorithms, it is expected that significant, although perhaps not generally applicable, computational simplifications can be found. That is, for any given combinatorial problem experience will indicate modifications which can greatly reduce computation time. This will eventually, it is hoped, permit the solution to larger problems in reasonable amounts of time.

Branching. We describe three heuristics for branching. Consider first the case where branching from a node N creates two new nodes $N + 1$ and $N + 2$. Node $N + 1$ has the property that $S(N + 1)$ contains those solutions in Ω which assign a given value to some variable, while node $N + 2$ represents the subset $S(N + 2)$ which contains those solutions in Ω which explicitly prohibit the same variable from taking on that value.

Within the above context two basically different alternatives for branching are possible. The first is to branch in such a way that $S(N + 1)$ is likely to contain the optimal solution, e.g., choose a variable and a value for that variable in such a way that $L(N + 1)$ equals a minimum. If this is done and branching is continued from node $N + 1$ the final nodes eventually created will contain, if not optimal solutions, solutions which are sufficiently good that they can be used for eliminating the need for exploring large numbers of other branches.

The second alternative would be to branch in such a way that $S(N + 2)$ is likely not to contain the optimal solution. In other words choose a variable and a value to assign to that variable in such a way that $L(N + 2)$ equals a maximum. The advantage of this is that nodes such as $N + 2$ will have high lower bounds, and therefore, once a final node is obtained, these nodes may not have to be further evaluated.

The second alternative is illustrated by the branching rule for the travelling salesman problem in which cities i and j are selected so as to increase most $L(N + 2)$. An example of the first alternative is to suppose that instead, cities i and j were selected as those which were nearest neighbors.

Finally, a third alternative may be available in which branching always creates m new nodes. Each of these nodes assigns to a variable one of m possible values. Thus, unlike the alternatives above, in this case, no nodes are explicitly created which contain subsets of Ω in which variables are constrained not to take on certain values. All nodes, here, represent subsets for which the variables have either assigned values or remain unassigned. An argument for this approach may be that nodes, when they are created, are all useful in that they assign values to variabl rather than prohibiting certain values. The job shop schedulin algorithm illustrates this type of branching rule. Recall, for th problem, branching creates m nodes where m is the number jobs not yet contained in the permutation being developed.

Bounding. The computational choice here is between effort obtaining high lower bounds for nodes as opposed to obtainin the same net effect by generating additional nodes.

Intuition would suggest that the higher up the problem tre (the highest node in the tree contains Ω; a node low in the tree therefore, is at the bottom of Figure 1), the more worthwhi may be the additional computation time necessary to obtai higher lower bounds. This is because of the proportionatel greater potential decrease in the number of nodes which need b evaluated should a sufficiently high lower bound be found. Thi is especially true if the stronger of two methods for obtainin lower bounds increases in computation time, only linearly wit problem size.

To illustrate this point using the travelling salesman problem; it may be computationally advantageous at some point before or during the branching process to solve an assignment problem using as a cost matrix $|c_{ij}|$. The constraints of an assignment problem are identical to those of the travelling salesman problem except for prohibiting the forming of disjoint sub-tours. Therefore, the assignment problem solution provides a lower bound on the travelling salesman problem solution. (In the fortunate case that the optimal assignment contains no sub-tours then the bounding procedure actually supplies the optimal solution.) Solving assignment problems for nodes near the top of the tree may permit discarding large branches from further exploration.

Choosing intermediate nodes. We discuss three alternatives available in deciding on an intermediate node from which next to branch. The first and apparently most popular choice seems to be from that node N for which $L(N)$ is a minimum. The rationale behind such a choice is that the minimum feasible solution is most likely contained in $S(N)$. A basic disadvantage of this rule is that the higher up the tree the farther will be $L(N)$ from ne minimum cost solution in $S(N)$. This is because there are wer variables for which values are assigned, and therefore, the lculated lower bound is generally weaker. Thus, the rule of oing to the intermediate node N for which $L(N)$ is a minimum as the property that it heavily favors nodes high up on the ee as opposed to what might be potentially more promising odes lower on the tree.

A second alternative is to continue branching from one of the n nodes created during the prior branching operation. If this hoice is used, branching would continue along the same branch ntil either all nodes became final or the lower bound of each xceeded the actual cost of a solution previously obtained. The dvantage of a procedure of this type is that certain information readily available when branching is from the node just created ut otherwise would need to be recomputed. In addition, as ointed out in Little *et al.* (1963) the procedure goes directly to a olution so that if calculations are stopped before optimality here is available a feasible solution as well as a lower bound on the optimal solution.

The third alternative is to do what might be described as choosing up the tree. The procedure is also described in Little *et al.* (1963). In this case branching always continues from one of the m nodes just created until eventually either final nodes are obtained, all nodes are infeasible or their lower bounds exceed the actual cost of a known solution. When this occurs the next intermediate node is chosen as follows. Trace back up the tree until a node, N, is found which has the property that one of the $m - 1$ other nodes created when branching took place from N is still an intermediate node. Branching is continued from this intermediate node. The advantage of this procedure is that it is necessary to retain information concerning very few nodes. Specifically, at any given time, only those nodes from which the best current solution was obtained and those on the branch being explored are required. On the other hand, the method may evaluate a relatively large number of nodes before an optimal solution can be obtained.

As an example of this process consider the node tree for a three machine, four job scheduling problem, as shown in Figure 2. The number in the nodes represents the partial sequences of jobs represented by each node. The node numbers are assigned in the order in which the nodes are generated. Nodes 9 and 10 are final nodes. To find the next intermediate node after nodes 9 and 10 are created, trace back up the tree to node 5 and choose node 7. When the nodes from $N = 7$ have all been fully explored,

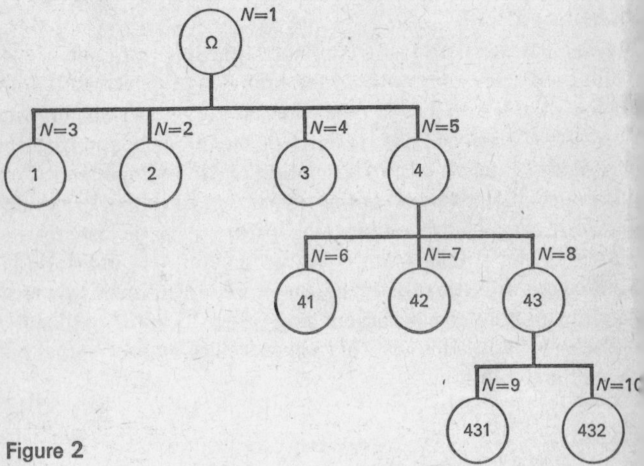

Figure 2

then the next intermediate node will be $N = 6$. After the nodes from $N = 6$ have been explored the intermediate node chosen will be $N = 3$. The procedure continues in this way until an optimal solution is obtained.

Conclusion

In concluding, two shortcomings of optimum seeking with branch and bound are briefly mentioned. First, although some of the terms and concepts associated with the technique have been generalized, obtaining optimal or near optimal solutions to any given problem still requires the determination of a specific set of rules and procedures. Each individual problem requires its own methodology for computing lower bounds and the determination of good heuristics for branching and choosing intermediate nodes. The problem here is not unlike that of choosing a variable to enter a current basis during linear programming iterations. One should not, however, be misled. There is a considerable amount of experience and folklore to be gained from an exposure to the solution methods of different problems and a knowledge of their computational characteristics.

Secondly, for large problems, it is quite likely that computational requirements will exceed the limits of available computer time. There is at least one partial remedy to this problem. The experience in solving linear programming problems is useful. When begun with good initial solutions, otherwise computationally excessive linear programming problems have been solved. The same, no doubt, holds true for combinatorial problems solved by branch and bound. In few cases has the idea been used.

Although branch and bound suffers from the above shortcomings, one decisive factor remains; that for a given problem it may provide the only known method of solution. It is on this basis that the value of the technique must be judged.

References

Balas, E. (1965), 'An additive alogrithm for solving linear programs with zero-one variables', *Operations Research*, vol. 13, pp. 517–46.

Brooks, G. H., and White, C. P. (1965), 'An algorithm for finding optimal and near optimal solutions to the production scheduling problem', *J. Indust. Eng.*, vol. 16, pp. 34–40.

Golomb, S. W., and Baumert, L. D. (1965), 'Backtrack programming', *J. Assoc. for Computing Machinery*, vol. 12, pp. 516–24.

Ignall, E., and Schrage, L. (1965), 'Application of the branch and bound technique to some flow-shop scheduling problems', *Operations Research*, vol. 13, pp. 400–12.

Land, A. H., and Doig, A. G. (1960), 'An automatic method of solving discrete programming problems', *Econometrica*, vol. 28, pp. 497–520.

Little, J. D. C., Murty, K. G., Sweeney, D. W., and Karel, C. (1963), 'An algorithm for the travelling salesman problem', *Operations Research*, vol. 11, pp. 972–89.

Lomnicki, Z. A. (1965), 'A branch and bound algorithm for the exact solution of the three machine scheduling problem', *Operational Research Q.*, vol. 16, pp. 89–100.

Further Reading

Books

J. Abadie (ed.), *Non-linear Programming*, North-Holland Publishing Co., 1967.

E. M. L. Beale, *Mathematical Programming in Practice*, Pitman, 1968.

R. E. Bellman, *Dynamic Programming*, Princeton University Press, 1957.

R. E. Bellman and S. E. Dreyfus, *Applied Dynamic Programming*, Princeton University Press, 1962.

A. Charnes and W. W. Cooper, *Management Models and Industrial Applications of Linear Programming*, vols. 1 and 2, Wiley, 1960.

G. B. Dantzig, *Linear Programming and Extensions*, Princeton University Press, 1963.

S. I. Gass, *Linear Programming*, McGraw-Hill, 1964 (2nd edn).

R. L. Graves and P. Wolfe (eds.), *Recent Advances in Mathematical Programming*, McGraw-Hill, 1963.

G. Hadley, *Linear Programming*, Addison-Wesley, 1962.

G. Hadley, *Non-linear and Dynamic Programming*, Addison-Wesley, 1964.

T. C. Koopmans (ed.), *Activity Analysis of Production and Allocation*, Wiley, 1951.

H. P. Künzi, W. Krelle and W. Oettli, *Non-Linear Programming*, Blaisdell, 1966.

W. Orchard-Hays, *Advanced Linear-Programming Computing Techniques*, McGraw-Hill, 1968.

V. Riley and S. I. Gass, *Bibliography on Linear Programming and Related Techniques*, Johns Hopkins Press, 1958.

S. Vajda, *Mathematical Programming*, Addison-Wesley, 1961.

S. Vajda, *Readings in Mathematical Programming*, Pitman, 1969.

Papers

E. Balas, 'An additive algorithm for solving linear programs with zero-one variables', *Operations Research*, vol. 13 (1965), pp. 517–46.

E. Balas, 'An infeasibility-pricing decomposition method for linear programs', *Operations Research*, vol. 14 (1966), pp. 847–73.

E. Balas, 'Discrete programming by the filter method', *Operations Research*, vol. 15 (1967), pp. 915–57.

E. Balas, 'Duality in discrete programming, 4: Applications', *Man. Sciences Research Report*, no. 145, October 1968, Carnegie-Mellon University.

E. BALAS and P. L. IVANESCU, 'On the generalised transportation problem', *Management Science*, vol. 11 (1964), pp. 188–202.

M. L. BALINSKI and R. E. GOMORY, 'A primal method for the assignment and transportation problems', *Management Science*, vol. 10 (1963), pp. 578–93.

M. L. BALINSKI, 'Integer programming: methods, uses, computation', *Management Science*, vol. 12 (1965), pp. 253–313.

E. M. L. BEALE, 'On quadratic programming', *Naval Res. Logistics Q.*, vol. 6 (1959), pp. 227–43.

J. H. BEEBE, C. S. BEIGHTLER and J. P. STARK, 'Stochastic optimisation of production planning', *Operations Research*, vol. 16 (1968), pp. 799–818.

J. M. BENNETT, 'Structured linear programming', *Operations Research*, vol. 14 (1966), pp. 636–45.

C. A. BLYTH and G. A. CROTHALL, 'A pilot programming model of New Zealand economic development', *Econometrica*, vol. 33 (1965), pp. 357–81.

A. P. G. BROWN and Z. A. LOMNICKI, 'Some applications of the "branch and bound" algorithm to the machine scheduling problem', *Operational Research Q.*, vol. 17 (1966), pp. 173–86.

D. J. CHAMBERS, 'Programming the allocation of funds subject to restrictions on reported results', *Operational Research Q.*, vol. 18 (1967), pp. 407–32.

A. CHARNES and W. W. COOPER, 'Chance constrained programming', *Management Science*, vol. 6 (1959), pp. 73–9.

A. CHARNES and W. W. COOPER, 'Programming with linear fractional functionals', *Naval Res. Logistics Q.*, vol. 9 (1962), pp. 181–6.

A. CHARNES, W. W. COOPER and G. L. THOMPSON, 'Constrained generalised medians and hypermedians as deterministic equivalents for two-stage linear programs under uncertainty,' *Management Science*, vol. 12 (1965), pp. 83–112.

A. CHARNES and M. KIRBY, 'Optimal decision rules for the E-model of chance constrained programming', *Cahiers du Centre d'Etudes de Recherche Operationelle*, vol. 8 (1966), no. 1, pp. 5–44.

G. B. DANTZIG, 'Linear programming under uncertainty', *Management Science*, vol. 1 (1955), pp. 197–206.

W. C. DAVIDON, 'Variance algorithm for minimisation', *Computer Journal*, vol. 10 (1968), pp. 406–410.

W. S. DORN, 'Non-linear programming – a survey', *Management Science*, vol. 9 (1963), pp. 171–208.

M. A. EFROYMSON and T. L. RAY, 'A branch-bound algorithm for plant location', *Operations Research*, vol. 14 (1966), pp. 367–8.

M. EL AGIZY, 'Two stage programming under uncertainty with discrete distribution function', *Operations Research*, vol. 15 (1967), pp. 55–70.

S. E. ELMAGHRABY, 'Allocation under uncertainty when the demand has continuous D.F.', *Management Science*, vol. 6 (1960), pp. 270–74.

S. E. ELMAGHRABY, 'The machine sequencing problem – review and extensions', *Naval Res. Logistics Q.*, vol. 15 (1968), pp. 205–32.

H. EVERETT III, 'Generalised Lagrange multiplier method for solving problems of optimum allocation of resources', *Operations Research*, vol. 11 (1963), pp. 399–417.

W. H. EVERS, 'A new model for stochastic linear programming', *Management Science*, vol. 13 (1966), pp. 680–93.

T. FABIAN, 'Blast furnace burdening and production planning – a linear programming example', *Management Science*, vol. 14 (1967), pp. B1–27.

A. V. FIACCO and G. P. MCCORMICK, 'The sequential unconstrained minimization technique for non-linear programming, a primal-dual method', *Management Science*, vol. 10 (1963), pp. 360–66.

A. V. FIACCO and G. P. MCCORMICK, 'Computational algorithm for the sequential unconstrained minimization technique for non-linear programming', *Management Science*, vol. 10 (1963), pp. 601–17.

A. V. FIACCO and G. P. MCCORMICK, 'The slacked unconstrained minimization technique for convex programming', *Soc. Indust. Appl. Maths Rev.*, vol. 15 (1967), pp. 505–15.

R. FLETCHER and M. J. D. POWELL, 'A rapidly convergent descent method for minimisation', *Computer Journal*, vol. 6 (1963), p. 163.

R. FLETCHER and C. M. REEVES, 'Function minimisation by conjugate gradients', *Computer Journal*, vol. 7 (1964), p. 149.

A. M. GEOFFRION, 'Reducing concave programs with some linear constraints', *Soc. Indust. Appl. Maths Rev.*, vol. 15 (1967), pp. 653–64.

A. M. GEOFFRION, 'Integer programming by implicit enumeration and Balas' method', *Soc. Indust. Appl. Maths Rev.*, vol. 15 (1967), pp. 178–90.

W. S. GERE, 'Heuristics in job-shop scheduling, ' *Management Science*, vol. 13 (1966), pp. 167–90.

P. C. GILMORE and R. E. GOMORY, 'A linear programming approach to the cutting stock problem', Part 1, *Operations Research*, vol. 9 (1961), pp. 849–59; Part 2, *Operations Research*, vol. 11 (1963), pp. 863–88.

P. C. GILMORE and R. E. GOMORY, 'Multi-stage cutting stock problems of two and more dimensions', *Operations Research*, vol. 13 (1965), pp. 94–120.

F. GLOVER, 'A new foundation for a simplified primal integer programming algorithm', *Operations Research*, vol. 16 (1968), pp. 727–40.

R. E. GOMORY, 'Some polyhedra related to combinatorial problems', *I.B.M. Research Report*, RC-2145, 25 July 1968.

H. O. HARTLEY and R. R. HOCKING, 'Convex programming by tangential approximation', *Management Science*, vol. 9 (1962), pp. 600–612.

B. B. HENRY and C. H. JONES, 'Linear programming for production allocation', *J. Indust. Eng.*, vol. 18 (1967), pp. 403–12.

R. A. HOWARD, 'Dynamic programming', *Management Science*, vol. 12 (1965), pp. 317–48.

M. KLEIN and R. R. KLIMPEL, 'Application of linearly constrained non-linear optimisation to plant location and sizing', *J. Indust. Eng.*, vol. 18 (1967), pp. 90–95.

K. O. KORTANEK, D. SODARO and A. L. SOYSTER, 'Multi-product production scheduling via extreme point properties of linear programming', *Naval Res. Logistics Q.*, vol. 15 (1968), pp. 287–300.

E. L. LAWLER and M. D. BELL, 'A method for solving discrete optimisation problems', *Operations Research*, vol. 14 (1966), pp. 1098–112.

E. L. LAWLER and D. E. WOOD, 'Branch and bound methods: a survey', *Operations Research*, vol. 14 (1966), pp. 699–719.

C. E. LEMKE and K. SPIELBERG, 'Direct search algorithms for 0–1 and mixed-integer programming', *Operations Research*, vol. 15 (1967), pp. 892–914.

J. D. C. LITTLE, K. G. MURTY, D. W. SWEENEY and C. KAREL, 'An algorithm for the travelling salesman problem', *Operations Research*, vol. 11 (1963), pp. 972–89.

B. A. MURTAGH and R. W. H. SARGENT, 'A constrained minimisation method with quadratic convergence', in R. Fletcher (ed.), *Proceedings of the BCS/IMA Conference on Optimisation, Keele*, 1965, Academic Press.

T. A. J. NICHOLSON, 'A boundary method for planar travelling salesman problems', *Operational Research Q.*, vol. 19 (1968), pp. 445–52.

R. L. O'MALLEY, S. E. ELMAGHRABY and J. W. JESKE, Jr, 'An operational system for smoothing batch-type production', *Management Science*, vol. 12 (1965), pp. B433–49.

C. C. PETERSON, 'Computational experience with variants of the Balas algorithm applied to the selection of R and D projects', *Management Science*, vol. 13 (1966), pp. 736–50.

J. K. SENGUPTA, G. TINTNER and C. MILLHAM, 'On some theorems of stochastic linear programming with applications', *Management Science*, vol. 10 (1963), pp. 143–59.

J. F. SHAPIRO, 'Dynamic programming algorithms for the integer programming problem', *Operations Research*, vol. 16 (1968), pp. 103–121.

J. F. SHAPIRO, 'Group theoretic algorithms for the integer programming problem', *Operations Research*, vol. 16 (1968), pp. 928–47.

W. F. SHARPE, 'A linear programming algorithm for mutual fund portfolio selection,' *Management Science*, vol. 13 (1966), pp. 499–510.

L. W. SWANSON and J. G. WOODRUFF, 'A sequential approach to the feedmix problem', *Operations Research*, vol. 12 (1964), pp. 89–109.

G. L. THOMPSON, M. FRED and S. ZIONTS, 'Technique for removing non-binding constraints and extraneous variables from linear programs', *Management Science*, vol. 12 (1965), pp. 588–608.

H. M. Wagner and J. S. C. Yuan, 'Algorithmic equivalence in linear fractional programming', *Management Science*, vol. 14 (1967), pp. 301–6.

H. M. Weingartner and D. N. Ness, 'Methods for the solution of the multi-dimensional 0–1 knapsack problem', *Operations Research*, vol. 15 (1967), pp. 83–103.

A. Yaspan, 'On finding a maximal assignment', *Operations Research*, vol. 14 (1966), pp. 646–51.

R. D. Young, 'A simplified primal (all integer) integer programming algorithm', *Operations Research*, vol. 16 (1968), pp. 750–80.

W. I. Zangwill, 'The convex simplex method', *Management Science*, vol. 14 (1967), pp. 221–38.

Acknowledgements

Permission to reproduce the readings in this volume is acknowledged from the following sources:

Reading 1	*The Journal of Industrial Engineering*
Reading 2	*Management Science*
Reading 3	*The Journal of Industrial Engineering*
Reading 4	*Operations Research*
Reading 5	The Econometric Society
Reading 6	The Royal Statistical Society and N. R. Paine
Reading 7	The Royal Statistical Society and E. M. L. Beale et al
Reading 8	*Management Science*
Reading 9	*The Journal of Industrial Engineering*
Reading 10	Operational Research Society Ltd
Reading 11	*The Journal of Industrial Engineering*
Reading 12	*Management Science*
Reading 13	North-Holland Publishing Company
Reading 14	*Operations Research*
Reading 15	*Operations Research*
Reading 16	The Royal Statistical Society
Reading 17	Operational Research Society Ltd
Reading 18	*Management Science*

Author Index

Subject Index